A HISTORY OF CHEMISTRY

A HISTORY OF
CHEMISTRY

Bernadette Bensaude-Vincent
and Isabelle Stengers

Translated by Deborah van Dam

HARVARD UNIVERSITY PRESS

Cambridge, Massachusetts
London, England
1996

Originally published as *Histoire de la chimie,* © 1993, Editions La Découverte.
All rights reserved.

Publication of this book has been aided by a grant from the French Ministry of Culture.

Library of Congress Cataloging-in-Publication Data

Bensaude-Vincent, Bernadette.
[Histoire de la chimie. English]
A history of chemistry / Bernadette Bensaude-Vincent and Isabelle Stengers ;
translated by Deborah van Dam.
p. cm.
Includes bibliographical references (p.   -   ) and index.
ISBN 0-674-39659-6 (alk. paper)
1. Chemistry—History. I. Stengers, Isabelle. II. Title.
QD11.B4413 1996
540'.9—dc20
96-26615

CONTENTS

A HISTORY OF CHEMISTRY

PROLOGUE

Often it is assumed that there is a history of chemistry, a history of physics, a history of each separate science. The division of learning into disciplines imposes itself as if by necessity. This seems quite natural to us because, in the cloistered world of scholarly disciplines shaped by the model of Auguste Comte's rigid classification scheme, we have always been served pre-packaged sciences, each one set apart from the others in splendid isolation.

But by adopting a particular distribution of disciplines too quickly, we risk overlooking fundamental problems, which are often the most interesting ones to ponder. The historian of science, by accepting the contemporary framework of disciplinary boundaries, tends to take for granted a structure that was pieced together with considerable effort in the past. Disciplines like physics and chemistry have not existed since the beginning of time; they have been built up little by little, and that does not happen without difficulties. Chemistry was not included among the ancient scholarly disciplines, but toward the middle of the eighteenth century it carved out an important place for itself in the academies and universities and among the enlightened public. In the nineteenth century it became a cutting-edge science, the very image of progress. How did chemistry become pre-eminent, and for that matter, how did it become a science?

Most histories of chemistry have given essentially the same answer to that question: chemistry became a science when it became disentangled from the mass of archaic practices and occult knowledge handed down by tradition. Its break with an obscure past of artisanal traditions and alchemy marked the beginning of its history. Opinions are divided on when the rupture occurred. Various authors, depending on their culture and country of origin, have placed it in the eighteenth century and chosen either Georg Ernst Stahl (1660–1734) or Antoine-Laurent Lavoisier (1743–1794) as the "father of modern chemistry." Others have preferred to go back to the seventeenth century and see the turning point in the work of Robert Boyle. But in every case the story of the past hinged on one or two fixed points at which the course of history changed direction. As if they had to find a "Newton" or "Galileo" at any price, historians have postulated the existence of a defining moment from which chemistry, finally awakened to its distinct identity, had only to proceed straight ahead to develop its scientific and technical potential.

Classical histories of chemistry were thus divided into two clear-cut periods: the prescientific age and the scientific age. Actually this view offers great narrative resources. Ferdinand Hoefer, for example, led his readers through a world of highly colorful tales (Hoefer, 1840–1842). First threading his way along the bosky paths of more or less magical practices, hermetical symbols, and exotic cultures, Hoefer soon reached the triumphal road of progress; once there his story became a "serious" history of a chemistry built on laws and experimental discoveries, the accumulation of which would naturally engender a host of industrial or agricultural applications of increasing benefit to humanity.

Today this kind of story seems a bit antiquated, a remnant of the serene and lofty profile that chemistry presented in the last century. It is a vestige of a time when chemists wrote their own history themselves. In the nineteenth century it was not unusual for a chemist who was making history with his writings and research to become a historian—sometimes an erudite one—and define the public image of his discipline. Thomas Thomson (1830–1831), Hermann Kopp (1843–1847), Adolphe Wurtz (1869), Albert Ladenburg (1879), Marcellin Berthelot (1890), Edward Thorpe (1902), Pierre Duhem (1902), Ida Freund (1904), and Wilhelm Ostwald (1906)—all wrote narratives of the past that served as manifestoes for their science. They described a chemistry that was sure of its identity—and of its successes, as well.

Narratives of this type are still popular today—books by François Jacob, Richard Feynman, and Ilya Prigogine are examples. But today's innovations in chemistry no longer stimulate a renewal of interest in history. It is as if chemistry's past can no longer be reactivated by its present. The history of chemistry is now written by professional historians, and it has been transformed in the process. The authoritative division into two periods—the prescientific and scientific ages—has not stood up to detailed analysis of archival materials, such as course notes, correspondence, manuscripts, notebooks, and laboratory instruments. Passing the work of famous scholars as well as that of obscure and anonymous chemists through the sieve of historical criticism, historians of science have eliminated much of the received wisdom that was promulgated in traditional histories and manuals of chemistry. Tranquil certitudes about the origin of chemistry, its date of birth, its nature and philosophy are all gone. The boundaries of the discipline have become more blurred, shifting, and permeable; formerly clear outlines have become strangely murky. Historiography has certainly illuminated and enriched our perception of specific events, but it has done so by sacrificing the global perspective of the evolution of chemistry. The great historical frescoes seem, if not discredited, at least given over to caricature.

Is it reasonable, then, to try to reconstruct a broad picture of the discipline, from the most distant past to the present? There is a risk that one who makes the attempt will fall victim to the illusion that somewhere there existed a well-defined territory of nature, called "chemistry," that was obscured by a fog of speculations and awaited the arrival of enlightened scientists to puzzle out its laws and its functioning. There is also a risk of getting lost in transit on the journey from origins hidden in the darkness of time and the most ancient myths to the present-day tangle of bizarre molecules, a universe of new materials on the cutting edge of technology.

This enterprise seems doomed to failure, to be drowned in a wave of perplexity and questions of methodology. Where should we begin? Did chemistry come into being when practical information began to be acquired and transmitted? In that case we have to start with prehistoric times, with the first techniques for making and controlling fire, the first processes for dyeing cloth and fermenting wine, and the first pharmacopoeias. Or does it begin with the first elements of rational knowledge? In that case it must start with the pre-Socratic philosophers and the earliest atomic and elemental theories. If our story begins with the link between theory and practice, alchemy must be the starting point. If a "history of chemistry" should be limited to that chemistry clearly identified as a science, the seventeenth century is the necessary origin.

The difficulty of choosing a starting point presages many other choices to come. Must the history of mining, dyes, glass, cosmetics, and medicine be included? If we accept at the outset today's definition of chemistry as the science of the transformations of matter, then all the alchemists, perfumers, metallurgists, philosophers, and dye makers who devoted their lives to transformations of one kind or another must be included in the "prehistory" of the discipline. But if we refuse to judge chemistry's past on the basis of its current definition, then we must decide who the central characters in this drama will be. With each step the historian stumbles over the difficulty of delimiting the subject.

In fact, overwhelming doubt offers an escape from skepticism, for all the obstacles to be met are reduced to the same gnawing question: what is chemistry? This central question, then, is our first clue for constructing the narrative. What if, instead of digging out the hidden past of a well-defined science whose identity is not in question, we envisage this science as the *product* of a history? What if, instead of saying that chemistry *has* a history, which one can choose to study or ignore, we propose that it *is* a history in progress? This history would less resemble the triumphal march of a science that is sure of itself than a long chain of events shaping a science that is haunted by questions of its nature. Is chemistry a science or an art? Is it discursive knowledge or a more or less coherent body of know-how? Does it form an autonomous system, or is it a set of doctrines whose rational center is to be found elsewhere? These are questions not only for historians but for chemists as well. By turns servant, mistress, or rival of its fellows, physics and biology, chemistry never stops redefining its identity and its place in the encyclopedia.

The Question of Identity

Thus it is the quest for the identity of chemistry that we propose to use as the leading thread of our narrative. Of all the sciences, chemistry exhibits, it seems to us, a peculiarity in the definition of its territory. Here is a body of knowledge with multiple facets and innumerable ramifications: it applies in the depths of the Earth as well as in space, and it is as important to agriculture as it is to heavy and fine industry and to pharmacy. Here is a science that spans the borders between the inert and the living, between the microscopic and the macroscopic. How can we assign an identity to a discipline that seems to be everywhere and nowhere at once?

This question sounds contemporary to us, but it has been asked, though framed in other terms, since the beginning. Chemistry has always been heir to a heterogeneous territory, one that defied all a priori definition and therefore challenged chemists to construct an identity for it. Chemists have continually been forced to defend the specific autonomy and rationality of their science, because their concepts and their methods formed nodes or crossroads among heterogeneous areas on the map of knowledge and because they held strategic but disputed places on that map.

History is far from neutral in this perpetual search for identity. This terribly old yet still new science, heir to the most ancient techniques that define humanity, nevertheless produces ultramodern materials. For centuries chemists have been trying to come to terms with their past, oscillating between the temptation to appeal to tradition in order to stave off attempts at annexation or to jettison the past like a heavy weight that must be thrown off so that their work would be considered science. At present they are silent. Now that twentieth-century chemists have lost their historical preoccupations, one sometimes encounters descriptions of chemistry as the place, by definition anonymous, where the solutions to immemorial needs are formulated: research on new materials, production of medicines, and so on. In other words, the interests of chemistry are confused with those of *Homo sapiens*. The history that we shall recount, however, is full of chemists' thoughts on their own identity.

It is therefore worthwhile to reactivate the traditional genre of writing history by going beyond local studies of a limited time period and attempting to paint broad historical pictures. Once it is admitted that there is no eternal essence of chemistry, no transcendent object that is unveiled over the course of the centuries, it is possible to understand all the industrial and intellectual adventures that at various periods have shaped chemistry's successive identities only by considering a global history of long duration. Whether they were successes or failures, these experiences acquire a different meaning when taken as a whole and viewed as a means of organizing chemistry into the subject matter of history. Just as Fernand Braudel, thanks to his long-term perspective, was able to "invent" through his histories of the Mediterranean a space that is both physical and human, though it was inaccessible to the people who lived that history, so the historian of chemistry can hope to identify the specific space of chemistry (Braudel, 1977).

This project inspires the overall organization of this book. Each of the five chapters presents a different face of chemistry, delineates its identity at a given time. The first, which attempts to settle the question of origins, presents the polymorphous variety of artisanal practices and cultural traditions from which, in the seventeenth century, grew that segment of the scholarly disciplines called chemistry. The second chapter, which covers the entire eighteenth century, reveals chemistry as a conqueror claiming the dignity and legitimacy of a science in many distinct ways. The third presents the academic and professional face of chemistry in the nineteenth century. By guiding us through some industrial landscapes of the nineteenth and twentieth centuries, Chapter 4 profiles chemistry in the world of work and production. Finally, Chapter 5 presents a chemistry whose territory has been increasingly dismembered into a multitude of more or less hybrid or autonomous subdisciplines.

For each portrait of the subject there is a corresponding picture of the chemists of that period. The characters put on view in the first chapter are alchemists, doctors, metallurgists, and mystics all at the same time, but they are also skeptics and rationalists. Their outstanding trait is their diversity, as the science that placed them retrospectively "at its origin" did not have the power to give them a collective identity. Most of the chemists discussed in Chapter 2, natural philosophers, doctors, or pharmacists by training, were either academics or lecturers who disseminated knowledge by conducting demonstrations in public. The third chapter is invaded by professors of chemistry and the fourth by chemist-entrepreneurs, lucky or unlucky inventors and engineers. Chapter 5 describes a new profession, that of the service chemist who works outside disciplinary boundaries and uses his or her chemical expertise in a variety of areas of research or production. The more professional chemists spread out into new venues, the stronger becomes the temptation to identify their methods and their science as the response to the immemorial needs of *Homo sapiens*.

Our object is to establish chemistry's position in the world of knowledge and culture through this succession of profiles of chemistry and chemists. The place of chemistry in the hierarchy of the sciences was always a matter of debate, and in each period its relationship with its neighbors—the physical sciences and the life sciences—had to be renegotiated. We will try to show that from the beginning chemistry's position in the geography of knowledge was determined by a combination of three factors that were constantly being redefined: laboratory techniques, professions, and institutions.

The objectives of chemistry are framed primarily in reference to a body of instrumental techniques. Without understanding these procedures and their evolution it is impossible to understand the doctrines of chemistry. Experimental practices can lead to the overthrow of a doctrine, but they can also change the norms or standards of explanation. We will see several examples of how experimental procedures—collection, weighing, and purification—caused the rules of demonstration and the criteria of validation to evolve. Nineteenth-century chemistry illustrates the tenets of a practical, experimental science extremely well, but that does not mean that the "true nature" of chemistry had finally emerged; it means, rather, that the epistemological dialogue among the chemists of the period had effectively legitimized their experimental practices. At least for a time.

The status of chemistry is also defined by relationships among those who call themselves "chemists," the creators of an autonomous science, and the artisans and later industrialists who transformed ways of life and means of production. Century after century, through a variety of formulations, the relationship between chemical science and industrial chemistry has been reshaped. For eighteenth-century chemists, the status of chemistry was interdependent with that of the methods and know-how of the artisans who inherited the tradition. While in the nineteenth century chemistry became the very model of a useful science, today it crystallizes people's fears of the harmful consequences of technical progress. But these divergent judgments all respect the categories of "pure" and "applied" science developed in reference to chemistry.

The scientific legitimacy of chemistry is finally decided at the level of institutions. The invention of printing, the academic rules of publication, the creation of scientific journals, the conventions of naming and units of measurement all set milestones in the history of chemistry. Little by little, these developments drew the line between a branch of knowledge retrospectively judged vulgar or occult and a science finally established as academic, respectable, and prestigious. The means of publication, formation, transmission, and popularization of chemical knowledge are not a subsidiary concern but contribute directly to the history of chemical doctrines.

The Choice of Objects

From what, exactly, is a history of chemistry made? Who will be the actors in this story? Chemists, no doubt. We will see a certain number of them parade by, some obscure and others famous, but always they have been chosen because they defined, even incarnated, one of chemistry's identities. Through their practices, their discoveries, or their teaching, at a certain moment they transformed perspectives, created or wiped out prohibitions, justified or extinguished hopes, confirmed or denied promises. We do not aim to produce a portrait gallery of individual geniuses who built the edifice of modern chemistry stone by stone. In describing Guillaume-François Rouelle (1703–1770), Claude-Louis Berthollet (1748–1822), Antoine-Laurent Lavoisier (1743–1794), Friedrich August Kekulé von Stradonitz (1829–1896), or Frederick Soddy (1877–1956), we will try to elucidate not only their positive contributions but also what they left out or left aside, and what they could neither understand nor conceive. Above all we want to stress the problems and programs without which their work, their ambitions, and their battles would lose all meaning.

Are we, then, going to produce a history of doctrines, theories, and concepts? Yes, because chemistry is not simply a collection of empirical recipes. As its most ardent defenders have emphasized, chemistry distinguishes itself by producing coherent theoretical systems. But we do not intend to present a catalogue of disembodied chemical doctrines extracted from their context.

Neither the chemists nor their doctrines will always have the leading role. Through the portrait of one chemist or another, we will try to evoke a community defined not only by its institutional affiliation but also by its shared belief in a body of theoretical concepts and by the use of common laboratory and language practices. In describing how individual chemists acquired the power to master and predict phenomena, we will emphasize the constant necessity of learning and negotiating. The true subjects of this history, in our view, are the various investigative practices—research strategies, experimental and mental tools—that chemists have devised, as well as laboratory instruments, materials, procedures, institutions, courses, and credits. We will describe a gamut, if not complete at least varied, of practices that organized the field of chemistry around projects or research programs.

In presenting research programs or traditions in this way, we will not lead the reader to believe that each period enjoyed a consensus of opinion and a homogeneity of practices. The situation is more complex than we suppose when we speak, for example, of the victory of one paradigm over another as pockets of resistance faded away. Resistance to conceptual, theoretical, or technical innovations is not a simple phenomenon of inertia to be resolved eventually by the passage of time. No chemical theory has imposed itself on its own, like a light obliterating the shadows. Behind each truth we discover campaigns to publicize and promote it, networks of alliances, and processes of selection. Thus history can revive theories frozen into laws or formulas by unveiling the historical and local circumstances in which the truths that form the foundations of the discipline were constructed. A doctrine's current functioning, its successes as well as its crises, can then be seen as the result of choices, of a series of decisions that were made at particular moments and that set the norms of research for one or several groups for a time.

In choosing to refer the successive identities of chemistry to investigative practices, we have been led to favor the cognitive aspects over the productive aspects of chemistry. While technical and industrial developments are integral parts of the evolution of chemistry, they also involve many other ingredients belonging either to the history of technology or to business history. Some historians describe innovations in terms of scientific push; others in terms of market pull, profit motives, and the like. Here, as in the history of science, contemporary historiography has shown both the inadequacies of such descriptions and the need to favor local circumstantial analyses over comfortable generalities. This growing emphasis on the particularities of technical and industrial history makes any continuous narrative of the type "pure chemistry versus applied chemistry" impossible. It puts us face to face with a multifaceted world in which the problem of identity, which we have chosen as our leading thread, is no longer of any help to us. We will therefore content ourselves with an incursion into the domain of the history of technology simply to discover what is specific to chemical production.

The fourth chapter will therefore describe in broad strokes, without entering into detailed analyses, the evolution of chemical production from the artisanal extraction of natural substances to the industrial production of the synthetic products that fill our lives today. Its object is not to present this evolution—or the technical prowess in innovation and the control of processes that it implies—as the simple result of advances in knowledge. We want to show the relationship between the triumphs of chemists in the world of agriculture and industry and the scientific, cultural, and social stature of chemistry in the nineteenth and twentieth centuries.

"Every chemistry student, faced by almost any treatise, should be aware that on one of those pages, perhaps in a single line, formula, or word, his future is written in indecipherable characters" (Levi, 1975, p. 225). This remark, by a chemist who is also a novelist, seems a useful guide for our foray into industrial chemistry. Just as each chemist bases his career on a formula or a molecule, each industry is based on a substance. Working with materials whose properties determined the methods and strategies of production, the chemical industry distinguished itself from other types of industry by its prominent material logic. At the beginning of the nineteenth century the production of soda, chlorine, sulfuric acid, and other basic products was based on salts; for agricultural chemistry nitrogen was the strategic element. In the second half of the nineteenth century the empire of fine chemistry was built on the structure of carbon.

If this material logic sheds light on the major episodes of expansion in the nineteenth-century chemical industry, it does not explain them completely. Given over to the dictates of profit or of war, twentieth-century chemistry would confuse matters even more. The two world wars mobilized all the chemical troops—chlorine, nitrogen, carbon—but afterward the idea of a strategic material became less and less relevant. Just as chemical science saw its territory dismembered little by little and its identity menaced, so the chemical industry in the twentieth century has tended to be dispersed into other sectors and to bend to their needs.

The chemistry lived by Primo Levi no doubt belongs to an outdated past, but if historians may be allowed to focus a moment on a profile that appears characteristic to them, we would choose the image suggested by his words. Linking the destinies of an individual and a molecule, chemistry defines very specific relationships between man and matter: neither domination nor submission, but a perpetual negotiation—through alliances or hand-to-hand struggles—among individual materials and human demands.

1

ORIGINS

The Heritage of Alexandria

Where shall we find the origins of this science that we call "chemistry"? If we use the term to refer to the techniques that we define as "chemical," we must look for their roots in prehistory. We speak of the "Bronze Age" and the "Iron Age," traditional categories that reflect, mainly, the fact that metal tools have resisted the effects of time better than other artifacts from the ancient world. Nevertheless, we have every reason to believe that metallurgical techniques had a great significance for those who practiced them.[1] We must also mention the arts of fermentation, dye making, and the production of glues, soaps, perfumes, medicines, salves, and cosmetics. The origin of the processes that we recognize as "chemical" truly is lost in the mists of time.

Doctrines

Must we go back to the doctrines that could be included in a "prescientific" chemistry? But how can we identify them? If it is a question of recognizing pronouncements concerning transformations of what we call "matter," the subject explodes. From Thales's famous dictum that "water is the fundamental principle of everything" to Aristotle's theory of the elements, the ideas that obsessed and still obsess chemistry—prin-

ciples, elements, atoms, the problem of differentiation, the relationship
between the one and the many, generation interpreted as an ephemeral
transgression of a static order or as the result of a perpetual conflict—are
already formulated and confront each other in the earliest writings.

That all matter consists of four elements united by love or separated by
hate, as Empedocles thought; that everything contains seeds in infinite
numbers, all of which are different from each other, as Anaxagoras believed; that the elements themselves must be understood on the basis of
their fundamental qualities—water cold and wet, earth cold and dry, air
hot and wet, fire hot and dry—as Aristotle insisted; is this chemistry? The
answer is yes if we point out that, among all the sciences, it is chemistry
that has inherited this style of explanation referring to a procreation, to a
production of diversity—in other words, a "physical" explanation in the
Greek (from *phusis*, meaning nature as a procreative power) sense of the
word. In that case we will have to agree that the modern sciences have been
badly named, since our chemistry is, much more than our physics, heir to
the problems of the ancient "physicists." In the eighteenth century many
of those that we call "physicists" today would have been called mathema-
ticians or natural philosophers. The "physicists" of those days were practitioners of the experimental arts (electricity and heat but also chemistry,
geology, and even medicine). On the other hand, the answer is no if we
insist that the "chemical" discourse be linked to the practical goal of
making sense of new and artificial transformations of matter.

Again, can we see in the ancient atoms of Democritus, Epicurus, or
Lucretius the "precursors" of our atoms? Yes, if we think that the atomist
doctrines oppose to the qualitative conception of generation a combina-
tory conception, and that this opposition will appear again in the heart of
eighteenth-century chemistry. No, if we look for ways in which those
ancient atoms guided, constrained, or inspired the practices of those who
worked on materials. Elements, principles, and atoms will accompany us
through the entire history of chemistry, but this does not mean that the
history of chemistry has been bound by a conceptual continuity. On the
other hand, the re-use of these terms will not be arbitrary: they express a
tension between the different ways of explaining qualities and their transformations that will be constantly redefined until the twentieth century.

Must we then return to the practices and doctrines that are called
"chemical" and are themselves lodged in a specific history? In that case
the question of origins becomes confused with the very name of the

modern science: according to one historian, *chemistry* derives from the Egyptian word for "black," which is itself named for the black earth of Egypt (Moore, 1939; Wojtkowiak, 1988). Others (Carusi, 1990) assert that the same word may derive from the Greek word *cheo*, which means to pour a liquid or cast a metal. Greek or Egyptian etymology? The question cannot be settled, because it sends us back to the great Hellenistic city of Alexandria, where it was already the subject not only of legends but also of speculations and discussions.

The Alexandrian Corpus

Alexandria, center of the mingling and re-creation of Greek traditions— Pythagorean, Platonic, Stoic, Egyptian or Asian, Gnostic—and heir to a past that she revered, is the origin of the specific character of *alchemy*. Alchemy's formidable riches, the power of its metaphors and analogies, still haunt our language, if no longer our ideas. Today we associate alchemy with magic, but originally it encompassed both mysticism and practical techniques for manipulating matter.

It was in Alexandria, crossroads of culture and commerce, that "chemistry" as a body of knowledge and experience devoted to retrieving a legacy, to deciphering, reconstructing, and transmitting a lost science, was born. Certain texts, such as the papyri of Leiden and Stockholm found in a tomb in Thebes, provide evidence of the transmission of artisanal recipes without mystical or philosophical pretensions. They include the description of a procedure, explicitly recognized as fraudulent, to give nonprecious metals the appearance of gold or silver.[2] On the other hand, they also include a work by Pseudo-Democritus,[3] the most ancient author cited in the surviving fragments of the Alexandrian corpus (as it was reassembled in the Byzantine period), entitled *Mystical and Physical Things,* which associated recipes for dyes and for making gold and silver with explanations inspired by the Greek theory of the elements and by astrology, all embellished with aphorisms that were to become the object of much speculation in the centuries to come.

The Hellenistic alchemists believed knowledge led to salvation, in the gnostic way. They saw themselves as heirs to the secrets of ancient Egypt, to the divine knowledge of Hermes Trismegistus.[4] Were they heir to the Chinese and Indian alchemical traditions as well? Although the themes are common—notably the link between an interest in metals and mystical themes going back to Taoism for the Chinese and to Tantrism for the Indians—the differences are important. For the Chinese, from the eighth century B.C. it was the secret of immortality that defined the quest. Interest in gold, which was thought to be inalterable and therefore immortal, was subordinate to this quest. The Chinese alchemists created "elixirs," based on mercury, sulfur, and arsenic, and attempted, without much success, to attenuate the toxicity of these poisonous brews. Joseph Needham has made a list of the emperors who may have died as a result of poisoning by elixirs (Needham, 1970). Chinese alchemy died out with the rise of Buddhism. As for the Indian alchemists, if they produced gold, they attached little importance to it. Their goal was not to "cure" metals but to "kill" them (to corrode them) in order to make medicines out of them.

The whole of the surviving Alexandrian corpus, whose best-represented author is Zosimus (fourth century A.D., the end of the Hellenistic period), provides a true cipher for the history that follows, the major references of traditional alchemy. In it we find the mythical "authors" of alchemy (Mary the Jewess, Agathodaimon, Cleopatra). We find the association between procedures, mystical symbolism, and cosmogonic doctrine, which is considered typical of alchemy. We also find the description of procedures (distillation, sublimation, filtration, dissolution, calcination, cupellation) that create a practical continuity between alchemy and chemistry.

Zosimus distinguished between "bodies" and "spirits," which could be extracted from bodies or bound to bodies. He described the sequence of colors that indicated the success of an operation, usually black, white, yellow, and purple. He professed to possess what would become the elixir[5] or the philosopher's stone, the tincture capable of accomplishing instantaneously what the alchemists had long worked on—producing gold, the ennobling or correction (cure) of "base" or "sick" metals—and he noted in passing that his tincture could also cure human ills. What in the Latin translation is called "to transmute," Zosimus called *baptizein*, that is, to immerse, as one immerses a cloth in a dye. That may mean that gold was defined first of all by its brilliant yellow color, a quality that would, no doubt, interest the dishonest jeweler. By the Middle Ages the answer to the question "what is gold?" involved much more rigorous tests, notably lack of corrosion upon exposure to air, resistance to cupellation (the process of melting gold on a bone cupel and blowing air on

its surface, which would oxidate lead if it were present) and to cementa-
tion (when gold was heated with a corrosive paste containing vitriol and
ammonia salts, it must not corrode at all). The term *tincture* would,
however, continue to be used.

From Arab Alchemy to Christian Alchemy

According to the scholar Ibn al-Nadim (tenth century), the first texts
translated from Greek into Arabic were texts on alchemy. These translations were made, no doubt, during the eighth century, in Damascus and
then in Baghdad, and it is through them that we have knowledge of the
Alexandrian corpus. Arabic alchemists began their work during the same
period. Jabir ibn Hayyan was the first of them (this name may be
fictitious, for we know nothing about him beyond the enormous body
of texts assembled under his name). This collection would continue to
grow, for many medieval authors would sign their own writings under
the name of "Geber," the European version of Jabir. Al-Razi, who lived
from 864 to 925, was the first alchemist whose work and life were described by reliable sources.

Arab Chemists and Alchemists

How is the revival of the "alchemical" question in the Arab world to be
understood? Since chemistry has been seen as the opposite of alchemy,
alchemy has traditionally been viewed as an erroneous doctrine that had
to be abandoned for truly scientific modern chemistry to become possible. Can we, however, identify the alchemical doctrine by its "erroneous"
character, thus pointing to the developments, innovations, and problems
rendered impossible by that doctrine? The question arises all the more
forcefully because we know the Arab world better than we know the
civilization of Alexandria. We can, therefore, better appreciate the vitality
of the "secular techniques" that developed there at the same time. What
was cause and what effect? Did the alchemical "dream" have technical
repercussions in the sense that, for example, sending men to the moon in
our own century resulted in innovations in "secular" (terrestrial) life? Or
did the Arabs become interested in the alchemical tradition because of
their interest in "chemical techniques" in the broad sense of the term? In

that case, alchemy would be distinguished rather artificially—for us and for the Arab thinkers who discussed its legitimacy—from a continuum of practices whose fruitfulness we must recognize.

In any case, Arab "chemistry" perfected the arts of distillation, the extraction of oils (perfumed essences), and the production of soap, metal alloys (the famous swords of Toledo), and pharmaceutical medicine. Glass, dyes, paper, colored inks—all were part of the refined and learned Muslim world. New or improved devices were invented: the hot water bath or "bain-marie" (by Mary the Jewess, one of the apocryphal authors of the Alexandrian tradition), hot air baths, perforated crucibles allowing separation as a result of melting, various types of retorts for distillation, sublimation, and other processes. The techniques are described with care and precision, the quantities of reagents and their degree of purity are specified, and signs indicating how to recognize the different stages are given. In brief, Arab scholars devoted themselves to the production and transmission of reproducible, practical knowledge, whether we call it secular chemistry or alchemy.

The classification of substances varies from one author to the next, but it generally relied on tests to which materials could be submitted or procedures that could be applied to them. "Test" must be understood here in a double sense, experimental and moral: gold was considered noble because it resisted fire, humidity, and being buried underground. Camphor, like sulfur, arsenic, mercury, and ammonia, belonged to the "spirits," because it was volatile. Glass belonged among the metals because, like them, it could be melted. And since the seven known metals—gold, silver, iron, copper, tin, lead, and mercury—were characterized by their capacity to be melted, what made a metal a metal was defined by reference to the only metal that was liquid at room temperature, mercury or quicksilver. But "common" mercury differed from the mercuric principle, which was cold and wet. Like all the other metals, it involved another "principle," which was hot and dry, sulfur.

As for the great work of transforming base metals into gold, the Arab alchemists looked to the authority of the Alexandrian scholars, who were thought to be heir to even more ancient knowledge. In monotheistic Islam, Hermes was no longer a god but a wise man, even a prophet. But from the tenth century onward the possibility of transmutation was the subject of debates and controversies. For Avicenna (980–1037) taxonomy was uppermost: the different metals belonged to distinct species and had distinct forms, and human art could not transform one into another. Classification according to species, which allows us to name things and put them in order, imposes limits that human power cannot overcome. On the other hand, from the alchemist's point of view, power was based on time, and order was temporal above all. Metals formed and matured slowly in the bowels of the earth, and the work of alchemy fit into this dynamic perspective: it was work in time, work that took time to accomplish. Alchemical procedures were long—counted in days rather than hours—but they attempted to reproduce in the laboratory, in the "artificial matrix" that is a hermetically sealed alembic (still), a process that takes centuries in nature.

The Christian World

Alchemy may be work in time, but it is also work within a tradition that is lost in the mists of time. Already in the Islamic world, the allegorical complexity of alchemical writings and the impossibility of knowing whether an author understood what he was writing or merely had copied a text that was obscure to him made alchemists the butt of criticism and mockery. The alchemical corpus was translated into Latin toward the middle of the twelfth century by Christian scholars divided between the noble desire to fight the infidel enemy and a devouring curiosity for their knowledge, and from the very beginning alchemy was a recognized but controversial science. Christian adversaries of alchemy did not have to invent polemical arguments against it, because denunciations had already been advanced within the Arab tradition. On the other hand, the intellectual, political, and theological stakes of alchemical doctrine, which emphasizes relationships among human powers, the transformation of matter, and the secrets of creation and salvation, no doubt gained a new intensity in the Christian world. For alchemy was entering a world in crisis, in which the development of urban centers and intellectual, commercial, and artisanal activities was blurring the distinctions between pagan knowledge and revealed knowledge, between the pursuit of salvation and practices of production, between faith and reason.

In *Les Origines sacrées des sciences modernes* [The sacred origins of modern science], Charles Morazé (1986) stresses the role played in the invention of new goals for alchemy by the newly created minor religious orders, the Dominicans and especially the Franciscans. The Dominicans

Albertus Magnus (1193–1280), Vincent de Beauvais (?–1264), and Thomas Aquinas (1228–1274) wrote about alchemy. (Thomas held that transmutation was a demonstrated truth.) The preoccupations of the Franciscans Roger Bacon (1214–1294) and Arnau de Villanova (1235–1313), as well as Raymond Lull (1235–1315), a mystic close to the Franciscans, cannot be separated from theological questions (the presence of the divine in the least being in nature), politics (the dignity of the poor and of manual labor), logic (anti-Aristotelian nominalism), and practices (purification, maceration—i.e., mortifying the flesh, rectification). All of these concerns reflect a questioning of the traditional opposition between the preoccupations of daily life and the order of salvation (Colnort-Bodet, 1986).

The imagery of Christian alchemy is well known. Saint George slaying the dragon, a king and queen entering a bath under threat of a naked sword brandished by a soldier, a wolf devouring a dead king, the fall of Troy and the death of Priam, aged Mercury simmering in a bath until his spirit, a white dove, escapes: each of these images is crowded with theological, mystic, mythic, astrological, and operational references (Morazé, 1986). The soldier was the solvent who would force the sulfur king and the mercury queen to react with each other; the death of Priam was the dissolution or melting of a compound; the hermetic dragon, who represented a test upon the alchemist's path, a counterforce opposing his work, guarded the grotto where Quintessence was found; the wolf was the antimony devouring the gold before the purifying fire regenerated a living, active, "philosopher" king; and so on. These references were enriched by themes belonging to Western Christianity. All the resources of Christianity were engaged in the invention of a Christian alchemy, which was, however, officially condemned by the ecclesiastical authorities: the mystery of the Trinity and that of the redemption, the conception of the Child God by Mary, Christ's passion, the search for the Holy Grail, the trials—contrition, mortification, purification—that marked the road to salvation, the many levels on which the spirit could play, from the Holy Ghost, whose sign was the dove, to the spirit of salt.

At the beginning of the fourteenth century, notably in the work of Raymond Lull, alchemical symbolism began to be elaborated in the form of geometric representations as well: the four elements of Aristotle, the four qualities, the successive operations of the *magnum opus*, the different metals, spirits, and salts were organized into rectangles, triangles,

stars, and circles so that the relationships between them could be studied
and the secrets of their numbers discovered.

The Question of Evaluation

Christian alchemy, like Arab alchemy, raises the question of evaluation.
The first problem is to decide who the medieval alchemist was. We must
stress, with contemporary historians of alchemy,[6] that medieval alchemy
must not be understood according to the better-known model of later
alchemy: an alchemy that coexisted with other, self-proclaimed rational
undertakings. Men of letters, who shared the opinion of Avicenna, could
be skeptical of the work of medieval alchemists. For example, Albertus
Magnus, the universal doctor, practiced alchemy, but he was doubtful of
the quality of gold produced by alchemists because nature was the only
alchemist. But he did not contrast alchemy to "another chemistry" believed to be rational.

Who was the medieval alchemist? Distinctions between those we
would call an artisan-chemist, a forger, a lucid and methodical researcher, a curious sage, an enlightened one, or a mystic are difficult to
make. Alchemy formed a continuum, and most authors remained humbly anonymous.

Only with the sixteenth century, with John Dee or Michael Maier,
does the continuum begin to become differentiated. The character of the
alchemist becomes recognizable, and disturbing links with societies that
are both secret and fascinating, such as the Rosicrucians made famous
by Umberto Eco in *Foucault's Pendulum,* are attributed to alchemists.
The familiar, modern image of alchemy as an occult science, as opposed
to the exact and soulless sciences, is the product of this differentiation.
It transforms the theme of secret knowledge—one cannot transmit to
just anyone a powerful secret that would be dangerous in the hands of
the unworthy—into the theme of intrinsically mysterious knowledge,
which may be revealed only after the student undergoes an initiation.
The similarly familiar argument that only a person who has been
purified—who has passed the tests of initiation that qualify him ethi-
cally—can accomplish the *magnum opus* is modern as well. This argument did not make much sense earlier, when the medieval alchemist had
to deal with the *testator* who, in the name of his financial backer, tested
what he said was gold. It made sense when the alchemist had to face a

skeptical group insisting that he demonstrate the reproducible character of his techniques.

But the question of evaluation creates another trap. When we insist on a history oriented toward modern chemistry, we evaluate the past in terms "in spite of" what we know today. In spite of their false belief in transmutation and the existence of a hidden knowledge, alchemists perfected the techniques of modern chemistry. They learned to produce ever more powerful acids (solvents). While the Arabs had only weak acids (solutions of corrosive salts), from the fourteenth century on European alchemists learned to prepare and condense strong acids: first nitric acid (*aqua fortis* or spirit of nitre), then hydrochloric acid (spirit of salt), then sulfuric acid (spirit of vitriol or, in its concentrated form, oil of vitriol), right up to *aqua regia* (a mixture of hydrochloric and nitric acids), which dissolved even gold. These acids were powerful instruments that made it possible to produce and characterize ever more varied salts and ever more numerous "spirits," the spiritual parts of the substances that distillation separates out. Finally, the alchemist, together with the *testator*, created an ever more exacting relationship between the identity of a substance and the tests it was capable of undergoing. In doing this they developed practices that gave a functional identity to substances that was based on the difference between the secondary properties (which could be imitated—that is, the properties did not define the substance) and properties that could be considered intrinsic. In other words, "in spite of" their false beliefs, alchemists prepared the field for the very practices that were to toll the death knell for alchemy.

But was this "field" only that of chemistry? Charles Morazé points out certain possible effects of the alchemists' geometric schemas: the quintessence (the "fifth essence"), put in the center of a diagram showing the four elements, where it represented the source of their transformations, implies the possibility of unifying the qualities that Aristotle believed contradictory to each other (Morazé, 1986, pp. 222, 268–269, 290–291). There would then be no logical opposition between cold and hot or between wet and dry, but only a quantitative difference subject to measurement and procedures. Alchemy would then be part of the conceptual path that led from Aristotle's science to the experimental sciences, which submitted qualities to measurement.

Finally, evaluating what alchemy achieved "in spite of itself" leads to the question of our own historiographical practice. As is often the case with the Middle Ages, the past is assessed according to the future, according to the discrimination of what time preserved because it was precious from what we see as mere waste. But what of this very mode of judgment? If we recall that the method of evaluation by separation and purification was an alchemical technique, if we recall such expressions as the "spirit" of a work, the "essence" to be extracted from a text, the "corrosive" value of a thought or witticism, or the "tests" that a thought must undergo to demonstrate its value, our views on historical evaluation become still more complicated. It's not just a question of metaphors. Indeed, our very intellectual attitude has been influenced by the alchemical tradition. The links woven by alchemy between time and value have been abandoned today, as far as material transformations (in which value is henceforth called rarity) are concerned, but those links remain for the "life of the spirit." How can we evaluate the meaning and importance of this "alchemy of thought," which seems to incline us toward time as a power, which judges, separates, purifies, matures, enriches and rectifies?

The second temporal dimension of alchemy, and the search for an ancient secret that lives within it, also brings us back to our own practices. The possibility of a *magnum opus* was the object of a thousand eyewitness accounts. Is it possible to doubt it? What evidence could counterbalance that of tradition? To what extent is this tradition reliable? These questions are part of our idea of rational criticism, but they are also questions that alchemy has given rise to throughout its history. From the beginning, alchemy was besieged by doubts about transmutation. The object of speculation about the possible and the impossible, the real and the fictitious, the site of disputes that pit the argument of authority against proof by empirical test, alchemy inspired invention and skepticism at the same time, mixed rational endeavor with speculation, and fed the idea of a secret to be discovered in nature as well as in books. In so doing it suggested a definition of intelligence as the answer to a challenge, the key to a secret that must be deciphered. Lying in wait for tokens, signs, and signatures, alchemy mobilized all the resources of the human spirit and applied them to the material world as well as to texts. Far from being perceived as a practice that prepared, "in spite of itself," the practical, intellectual, and emotional innovation that we associate with modern times, should alchemy not then be recognized as one of its main sources?

A Tradition in Crisis

In the contemporary history of science, "crisis" refers to the confrontation, described by Thomas Kuhn, between two rival theories. To a first approximation, one can say that "modern physics" was borne of such a confrontation. In any case, it was under the rubric of crisis that Galileo chose to place his work.

In the history of chemistry there is no equivalent of a "Galilean revolution": there was no one scientist who not only claimed to initiate a radical difference between the prescientific past and the scientific future but who also succeeded in having this claim recognized up to the present day. The identification of alchemy with an erroneous doctrine should lead us to think that the question of the possibility of transmutation—can man create in the laboratory what occurs naturally in the depths of the earth?—was the turning point in the course of history that led from alchemy to chemistry. But this is not the case, as we shall see. Explanations for the transformation of matter were not a crucial point, either. At the beginning of the eighteenth century the great chemist Georg Ernst Stahl would speak of "affinities" in a sense that alchemists since Albertus Magnus would not have rejected: like attracts and unites with like, so the dissolving of metals by acids, which manifests their affinity, proves that they share the same principle.

Modes of Transmission

We therefore propose a non-Kuhnian hypothesis: the difference between those who call themselves chemists and those who call themselves alchemists is linked to the question of the status of the discipline in question and to its manner of transmission. The role of printing in creating the crisis in alchemy is thus crucial. The time was past when an author, humbly or prudently, put the name of the famous Geber (Jabir ibn Hayyan) on his manuscripts and neglected to date them, a practice that created an obscure and migratory population of translations, commentaries, interpretations, and descriptions of new or ancient procedures. Printing forced authors to announce themselves as such, to take a position on the source of their authority, to address themselves to a certain type of reader, to subscribe to a type of history. Printing also transformed their relationship to authority. As soon as ancient texts were published, they became accessible to a wider public, and they could be confronted and challenged by modern authors: "When authors of atlases and herbals called on their readers to send in notes about coastlines or dried plants and seeds, a form of data collection was launched in which 'everyman' could play a supporting role" (Eisenstein, 1983, pp. 235–236).

Galileo is known to have contrasted the verdict of experience with the authority of Aristotle. In the sixteenth and seventeenth centuries, a number of authors asserted the right to cite the "book of nature" against scholarly authority—and this book was open to all: the educated, the curious, and the artisans.

Much more than this or that point of doctrine, the split between the idea of authorship and the idea of authority and the possibility of transmitting information and procedures help us to discriminate between those who would henceforth be called chemists and those who would pursue the mystery of alchemy. Esoteric transmission would face off against a didactic undertaking addressed to the public rather than to the initiated. After Leonardo da Vinci, Bernard Palissy (1510–1590), Benvenuto Cellini (1500–1571), and Georg Bauer, called Agricola (1494–1555), renounced secrecy and attempted to describe their techniques in a precise way so that their results could be reproduced. Their works are usually placed in the rather modest category of "technical chemistry" or "practical chemistry," because their creators remained outside the quarrels of other chemists. But Agricola's *De re metallica*, the sum of all the known metallurgical procedures, would remain a respected reference for a long time. Palissy and Cellini were technicians and artists at the same time: the ceramics and enamels of the first and the colossal bronze statues of the second were the result of systematic empirical research bearing not on the rediscovery of old secrets but on the development of new techniques.

Others, such as John Dee, Michael Maier, and Heinrich Cornelius Agrippa (who, before he died, denounced the illusory character of all the occult sciences that he had practiced), would stress the secretive, mystical, and magical dimensions of alchemy. But the secrets of alchemy changed in meaning when they were no longer transmitted only in the discrete form of copied and recopied manuscripts. A traditional network in which the initiated communicated with each other by

means of manuscripts was maintained; the "true" alchemists, like Newton, continued to copy and to write without intending to publish their
work (see Dobbs, 1975). But a large number of the old manuscripts
was published, along with a flurry of new works. Once printed, alchemical literature became public, even if the public was warned that
it would understand nothing of it. Its descriptions, reputed to be impenetrable to the uninitiated, became a "public secret." Thus, when
Ben Jonson used Michael Maier as his target in his famous play, *The
Alchemist* (1610), he was able to render the style and vocabulary of
alchemy perfectly.

Publishing alchemical procedures allowed anyone—such as a chemist
who had not made the effort to join a network of alchemists but had
simply bought a book instead—to put the techniques to the test and
expose errors. Here we see the strategic value of the theme of the relationship between the alchemist's spiritual purification and his power
over materials: it allowed him to state positively that his results could not
be reproduced by ordinary, lowly chemists. Alchemists would thus become what they remain today: objects of fascination and derision.

Revolutions

The best known of the sixteenth-century chemists is undoubtedly
Paracelsus, Theophrastus Bombastus von Hohenheim (1493–1541).
Should he be considered one of the magi of alchemy? For Paracelsus, all
those who improved upon nature—the baker who turned grain into the
perfection of bread, the metallurgist who transformed metallic ores into
swords, the farmer who created wine out of grapes—could be called
alchemists. God, who created the world, was an alchemist; the body,
which digests and transforms food into the human body, the noblest of
all bodies, was an alchemist. The creation of a homunculus in a flask
from spermatic liquid was alchemical as well. And above all the physician who could treat the body's ills was an alchemist. Man, as the center
of creation, possessed in himself a knowledge of things, but this knowledge could become reality only through experience, through the sympathy, attraction, and affinity between these things and their analogue in
man. Human beings could realize this knowledge only through the grace
personally conferred on the researcher by God. Empiricism and mysticism played off each other (Hannaway, 1975, pp. 25–26).

Paracelsus created a new fashion craze, a "popular" alchemy that was spectacular and public, that excited enthusiasm, passions, controversies, and hostility. He wrote in German in a provocative, extravagant, and pompous[7] style, insulted the ancients, and burned Avicenna's books in public (a demonstration suitable for the age of Luther, but also the age of printing). He declared that he preferred the science learned among the humble during the course of his wandering life to that of Aristotle and the learned authorities of the universities. He called upon doctors to learn "chemistry" and announced miraculous cures, such as the use of salt of mercury to treat syphilis. He diagnosed the external source of certain maladies, such as miners' disease (silicosis).

Even though Paracelsus rejected Aristotle's four elements and accepted the primary bodies of the alchemical tradition—mercury and sulfur plus salt, which was the element responsible for the passive solidity of things[8]—Paracelsian alchemy no longer turned toward tradition but proclaimed itself a "revolutionary," conquering science. Long after the frenzied struggle between Paracelsian chemical medicine and traditional Galenic medicine, Paracelsus remained in the memory of chemists (see Debus, 1992). Even in the eighteenth century the chemist Venel, in the article "Chymie" in Diderot's *Encyclopédie,* wished for a "new Paracelsus" to raise chemistry to the rank that she deserved among the sciences.

The passage from alchemy to chemistry was thus not marked by a theoretical break, such as the end of the belief in the possibility of transmutation or the rejection of alchemical and mystical references. Paracelsus can be called a chemist to the extent that he opposed to the authority of doctrines what he considered to be the authority of experience and saw himself as an innovator in a future-oriented field. On the other hand, his critic Andreas Libavius favored a return to Aristotle (see Hannaway, 1975). In his *Alchemia,* which appeared in 1597 and was devoted to a critique of Paracelsian chemistry, Libavius attacked the philosophico-mystical pretensions of the Paracelsians, denounced the antireligious, idolatrous, satanic character of the Paracelsian sects, and transformed the practice of chemistry into a simple illustration of the traditional dialectic that was taught in the universities in the name of Aristotle. He thus opposed to empirico-mystical Paracelsian medicine a didactic chemistry that was demonstrated through successive definitions, each definition generating the following ones by dichotomous division. The object of this demonstration was to classify all the procedures of an artisan's repertoire within such a plan, but the plan itself was not an organizing principle. Certain synonymous activities were distinguished from one another, as the artisans themselves had traditionally done.

Although Libavius returned to the traditional scholastic framework, he did start a new venture: he attempted to organize chemical knowledge in a rational way.[9] Hannaway (1975) stresses the "self-generating" power of a didactic strategy: as soon as a first classification is recognized, it *presents a problem,* and others attempt to solve the problem. Libavius thus initiated a process of confrontation between organizing hypotheses and traditional empirical procedures. There is no equivalent to this process in physics. Galileo, Newton, and their successors won the freedom to "create" or discover the laws of nature, which would become the object of their theory, and to display in those favorable cases the power of their ideas. The equivalent does not exist in biology either, for the biologist deals with the dense diversity of living beings but not with the tangled labyrinth of the procedures, properties, and substances with which the chemist is confronted. What the chemist knows about substances and their transformations cannot be separated from his techniques, and his techniques cannot be seen as logically deducible from a well-founded theoretical hypothesis. To the contrary, the field of chemistry reflects the practical usefulness of abandoned doctrines and hypotheses.

Alchemical Inventions

The alchemical doctrine was, indeed, fruitful in the hands of Jan Baptista van Helmont (1577–1640), a chemist, medical doctor, and disciple of Paracelsus. Helmont adhered to the alchemical vision that the generation of minerals is analogous to that of living things, and he attempted to define the seeds of this process. Unlike his predecessors, however, he rejected the theory of elements as well as that of primary bodies. Instead, he believed water to be the unique primordial substance: were plants not capable of growing in water alone? But just as there are two sexes, there must be two sources for the natural development of minerals and metals: water and the "seed," or "seminal spirit." The seed responsible for creating the shapes of substances cannot be identified with any particular substance. The question, then, is: through what material intermediaries, or by what instruments, does this seed—or the *archeus*, the agent postulated by Paracelsus to give form and unity to natural bodies—operate?

This doctrine inspired Helmont to make the first study of what, using his terminology, we call "gas," which he distinguished from easily liquefied vapors.[10] In particular he identified "spiritus sylvestris" (our carbon dioxide), which emerges when a natural substance, which has a shape, is destroyed by burning, as when acid is poured on limestone or during fermentation. Helmont further showed that this gas does not support combustion. So in the middle of the seventeenth century an alchemical doctrine inspired a new experimental application, creating new problems and necessitating the development of new techniques: "bell jars" used to receive the gases given off by reactions and to identify them would become part of laboratory equipment.

Even in the seventeenth century, Johann Rudolph Glauber (1604–1670) and Johann Joachim Becher (1635–1682) pursued a speculative, alchemical tradition. Glauber called the sodium sulfate that he prepared, among many other medicines, *sal mirabile*. He kept some of his techniques secret, not for the sake of alchemical mystery but for commercial property rights: he sold his medicines for payment. And he saw in the development of a national chemical industry a way to ensure the "health of Germany," which had been devastated in the Thirty Years War. Becher tried to persuade a prince, Herman von Baden, to invest in gold making, but he was equally involved with "cameralist" (economics-based) campaigns for the application of chemical techniques to a new area, that of the "public welfare" (Meinel, 1983). A large part of Becher's work was devoted to big projects: the digging of a Rhine-Danube canal, the administration of colonies in South America, and industrial developments (notably industrial uses for coal, tar, and the combustible gas obtained by distilling coal).

For the Kuhnian notion of a crisis caused by the clash of two paradigms, we must substitute here the picture of an indecisive struggle between rival doctrines. Alchemical magi, heirs to mysterious traditions, and the conquering alchemists, with their eyes fixed on the future, all lived in the same world with the Aristotelian "rationalistic" chemists, the chemist/technicians, both artists and artisans, and doctors, pharmacists, and metallurgists. There is no criterion that allows us to choose which among them would be the "bearers of the future."

A Science of Mixts or Corpuscles?

In 1417 Lucretius's *De rerum natura* was rediscovered. The story of the popularity of atomic or corpuscular theories associated with this rediscovery does not properly belong to the history of chemistry. The distinction between the primary qualities or properties of atoms and the secondary qualities, which depend on the observer, was the focus of a multitude of debates in philosophy, theology, physics, and esthetics. Aristotle had opposed atomism, and since antiquity it had been held to be an eminently suspect doctrine. In Christian Europe, too, Lucretius was considered an impious, materialistic author, and his disciples were suspected of atheism. When, in 1624, two Parisian scholars, Jean Bitaud and Antoine de Villon, put the doctrines of Aristotle and Paracelsus to a public trial, proposing to prove by experiment that everything is composed of indivisible atoms, the meeting was dispersed by the police and the participants were arrested and commanded on pain of death to promulgate their doctrine no more (Meinel, 1988). If atomism presented problems for theology, however, it was also able to recast itself as the only authentically Christian doctrine, as opposed to pagan Aristotelianism. Is it not the case that Aristotelian substances act according to their own nature, and do they not therefore enjoy a dangerous autonomy from divine power? In the seventeenth century, Pierre Gassendi in France and Robert Boyle in England argued that the existence of atoms meant that the world was like a machine, subject to the will of its Creator, to whom all glory was due.

Although the interest in atomism was a global cultural and intellectual phenomenon, specific relationships would be created between atomist doctrines and the preoccupations of chemists. In the seventeenth century chemistry became the favorite field for debate on atomism. Among the "empirical facts" cited to support the existence of atoms, chemical procedures held a central place: evaporation, rarefaction, condensation, and dissolution could all be described in terms of discontinuous matter. Furthermore, chemistry was also a privileged field for the confrontation between atomism and Aristotelian doctrine. Although the new mechanical science was, by definition, anti-Aristotelian, chemistry involved the problem of chemical transformation, and this problem puts into question what atoms are, what they can explain, and what an atomist explanation implies.

Atoms and Mixts

Two chemist-doctors, Daniel Sennert of Germany in 1631 and Sebastien
Basso of France in 1636, faced the problem of the scope and implications
of a "corpuscular chemistry": What is the relationship between *atoms*, the
ultimate particles, and *corpuscles*, the "actors" that carry specific chemical
properties and can be identified by means of chemical operations?

Basso and Sennert, unlike the alchemists and chemists mentioned so
far, were what could be called "ideologues" (though the word did not
exist then). Their real partners were not chemists but scholars, in the
broad sense of the term: thinkers who were either critical or respectful
of Aristotle and devoted to conceptual distinctions. The practices of
chemists essentially escaped discussion in these circles. The corpuscular
chemists were not interested in the number of primary bodies or the
criteria for defining them. All those criteria, in effect, referred to "functional properties," such as solubility and insolubility, combustibility and
volatility, which are made manifest during operations and make it possible to explain them. Corpuscular chemistry, in contrast, decoupled
concepts from operations; in other words, it provided explanations in
which concepts were relevant to any chemical compound indiscriminately, independent of the production processes or tests that established
its identity.

The crux of the dispute between the atomists and the disciples of
Aristotle was the question of the *actual* or only *potential* existence of the
constituent elements of a compound, which Aristotle had called *mixis*
(*mixtio* in Latin): what was the nature of elements when they entered
into the composition of this or that substance (Dijksterhuis, 1950, p. 74)?
The Aristotelian notion of the mixt assumed that every substance was a
compound and that the composition of a substance involved an internal
transformation of the components. In this sense the Aristotelian mixt
was already an antiatomic idea: unlike Democritus, Aristotle stated that
the *stoicheia*, the constituent elements of a body, were transformed
during the process of composition. He also asserted, however, that the
dynamis of the elements remained in the mixt; in other words, in one
way or another, the properties of a mixt reflected the properties of its
constituent elements.

The mixt was the focus of many discussions in the Middle Ages. It was
in effect a favorite subject of conceptual experimentation, in which the

relationship between form and matter could be analyzed. When a chemical reaction produces a new substance with new properties, we must conclude that this substance has a "substantial form" of its own. Where does it come from? And when a substance is destroyed, in the sense that it becomes part of a new compound, does that mean that its form is destroyed or that it continues to exist in a potential way, or in a weakened way, dominated by the new substantial form of the mixt? Avicenna thought that the forms of the elements remained but that their qualities underwent a *remissio,* a weakening of their intensity; Averroës suggested a *remissio* even of the form of the components; Thomas Aquinas suggested the destruction of these forms, the qualities of the constituents being integrated into a *qualitas media,* a middle quality, which rendered the compound susceptible to receive the *forma mixti,* the new substantial form. These discussions became even more vital because the question of the Eucharist was being argued out: in what sense did the bread and wine become the body and blood of Christ? In any case, the Aristotelian theory of mixts, in spite of its difficulties, made it possible to assert that the essential part of the dogma, "this is my body," could be taken in the strong sense, in the sense in which the bread took on a new substantial form. The fact that during the Council of Trent (1547–1563) the dogma of transubstantiation was enunciated within the conceptual framework of the theory of mixts would be an obstacle to atomism in Catholic countries.[11]

Corpuscular chemistry, as a site of conceptual experimentation on the consequences of atomism, assumed that the constituent elements did not continue to exist "potentially" in a compound but composed it *actually.* Chemical transformation had to be thought of in terms of the separation and combination of particles which were themselves invariant and incorruptible, existing prior to combination and remaining themselves in mixts.

The idea of chemical combination had, from the beginning, a great polemical value. It ran counter to the usual interpretation that alchemists and other metallurgists gave for the formation of metals: the production of bodies by maturation, by means of heat (the dry method) or by solvents (the wet method). It also made nonsense of all the discussions on substantial alterations that occupied Aristotle's disciples. From the atomist point of view, the central concept of Aristotelian science, the passage from potentiality to actuality, no longer made sense. The constituents of this world were actual. Correspondingly, genesis *(genesis)*, destruction *(phthora)*, and alteration *(alloiosis)* no longer referred to qualitatively different processes but to the kind of quantitative change that Aristotle called locomotion *(phora)*. In other words, the idea of chemical combination excluded from chemistry the ancient "physical" *(phusis)* or dynamic dimension of matter, and with it the time necessary for transformations. The living being needed time to arrive at the adult stage of life, but combination could occur instantaneously: time mattered only when obstacles must be overcome.

Because *separation* and *combination* are terms that we understand without difficulty, we may be tempted to see in corpuscular chemistry the stable point of departure for modern chemistry and therefore to go on to the next chapter. To do so would be to decide in favor of an argument because it is familiar to us; but, as we shall see, this argument had not, at that time, acquired the actual means of being right.

The Chemical Alphabet

Sebastien Basso compared atoms to alphabetical characters or building blocks. In both cases the comparison required that one distinguish between the properties of the mixt (the meaning of the text or the style of the building) and that of its components. We must note, however, that the two comparisons point toward two different conceptions of the atom. The "alphabetical" atom, borrowed from Greek tradition, was differentiated, even if it did not have the qualities of a text. Because each atom was a different "letter," they made sense when they were put together. On the other hand, bricks are all alike, even if the buildings they make up do not necessarily resemble each other at all. The "brick" model of the atom went back to a primordial, homogeneous matter and reduced qualitative differences to a question of configuration. Chemists argued over and oscillated between these two metaphors.

Moreover, was it enough to assert that a text consists of letters to reduce the text to letters? Sennert hesitated: as a chemist he considered it essential to assert that a compound is an aggregate, but as a philosopher he could not help asking about the identity of the aggregate. He ended up taking a position close to that of Avicenna, who had been criticized by the scholastic thinkers for taking a stand too close to a corpuscular philosophy.

And, finally, what was the relationship between the atoms and corpuscles to which a chemical identity could be attributed? From Sennert's atomic point of view, units of experimentally manipulable material were themselves heterogeneous aggregates, *prima mixta*. Basso, on the other hand, introduced the concept of corpuscles distinct from secondary, tertiary, and quarternary aggregates that may seem to foreshadow the modern distinction between atom, molecule, and compound but that actually reflects the indeterminacy of the connection between the idea of the atom and practical chemistry. In particular, the atomic doctrine in no way excluded the possibility of transmutation. Quite to the contrary, it eliminated the obstacle that was presented, especially for Avicenna, by the idea of substance inherited from Aristotle: if lead and gold were distinct, substantial forms, the transformation of one into the other seemed impossible; if they were different aggregations of atoms, it became quite conceivable.

Corpuscular chemistry was therefore less an answer than a new problematic framework in the double sense of "allowing the statement of a problem" and "creating a problem." While atomism could be associated with mechanics without difficulty, because it favored displacement among all the modes of change, corpuscular chemistry called chemical tradition into question.

Atomism, as the metaphors of the alphabet and the bricks indicated, seemed to promise a science constructed on a solid foundation: the atom was a principle of the construction both of reality and of knowledge. But it obscured the principle of the activity of the ancient alchemist and of the chemist: the possibility of qualifying a body, or the fact, as would be said in the eighteenth century, that a chemical transformation creates *this or that* homogeneous substance from heterogeneity. What *made* for homogeneity? The concept of the "mixt" reemerged in eighteenth-century chemistry, and this comeback was not due to any "regression" or to the persistent influence of Aristotle. In fact, Aristotle himself had been confronted by the problem of the mixt, of the production of a new body from other bodies with different properties. From the point of view of chemical operations, however, the new model of combinations and separations of invariable particles had another effect. It tended to favor those chemical procedures that were *reversible*.[12] For example, gold apparently dissolved in *aqua regia,* but it could be recovered later on. For Daniel Sennert, all the known empirical processes through which a metal could be separated—reduced to its primitive state *(reductio in pristinum statum)*—were proofs of the existence of atoms. This proof was not sufficient, the idea of the mixt being perfectly capable of explaining reversibility, but the operations that brought reversibility to the fore inspired a new classification of chemical procedures that theoretically favored purification processes. That which disappeared and then recomposed itself could be designated a "pure" body: a body whose identity was independent of the transformations one submitted it to or the source from which it was extracted.

At its origins, corpuscular chemistry was essentially a new conceptual theory, incapable of guiding the work of the chemist in the laboratory. However, the time required for reactions characteristic of the atomist theory—the time needed for making or dismantling aggregates—conferred on it the power to compete with the time of the maturation processes implied in the tradition of mixts.

The Atom without Qualities

With its distinction between primary properties (extension, form, impenetrability, mass) and secondary ones (color, heat, sound, etc.), the mechanist version of atomism denied all qualitative differences to atoms and granted them only geometrical attributes. It was able to attract mechanists like Galileo and philosophers like Descartes and Locke, but it increased even more the perplexity of the chemist who tried to construct a general theory of the principles of chemistry on the basis of this atom "without qualities." Differentiation between "secondary" qualities and those that allow us to define a body is not new; it is a constituent problem of the profession of chemist. If chemistry has retained something of the alchemical tradition, it is that "all that glisters is not gold." On the other hand, the "primary" properties that mechanist atomism offered the chemist might allow him to interpret the important difference between stable bodies and volatile ones and between dissolution and evaporation. But those properties left undetermined the "purely chemical" properties that defined the capacity of one body to destroy another, the tests a body could resist and those that destroyed it.

Historians gladly use the category of "mechanist chemist" to denote those who attribute only primary properties to atoms—in other words, those who opted for the metaphor of the brick rather than that of the alphabet letter. But what did the chemists Robert Boyle (1627–1691) and Nicolas Lémery (1645–1715), who both made this choice, have in common?

A Cartesian Novel

How are we to explain the acidity of a liquor in Nicolas Lémery's perspective? He thought an acid contained "pointed" particles, as shown by the tingling on the tongue and the shapes of crystallized acid salts. The power of an acid depended on the sharpness of the points of its particles, on the ability of the particles to penetrate the pores of the bodies they attacked. Limestone fizzed when it was put in contact with an acid because it consisted of rigid and brittle particles: the acid points penetrated the pores of the limestone, broke them up, and carried off anything that restricted their movement.

Lémery constructed a Cartesian romance in which the only characters were shapes and movement (Metzger, [1923] 1969, pp. 281–340; and Duhem, [1902] 1985, pp. 17–28). An unexpected feature of his approach was that chemical transformation was portrayed as violent: in Lémery's view it was not a question of affinity, attraction, and sympathy but of combat and destruction. Acidity was lost during a reaction because of the destruction of the particles' points.

The *Cours de chymie* that Lémery published in 1675 was an enormous success: according to Bernard le Bovier de Fontenelle, it was a "completely new science that came to light and that excited our curiosity" (quoted in Duhem, [1902] 1985, p. 23), a science finally released from all reference to occult qualities, from all barbarous and murky jargon. It enjoyed, again according to Fontenelle, a wide circulation, like gallant poetry or satire. It was translated into Latin, German, English, Spanish, and Italian. And in the cellar of the Rue Galande a large and varied audience of ladies and students attended Lémery's lectures for twenty-five years. But his chemistry was new only in its recourse to Cartesian imagery. He laid out the empirical knowledge of seventeenth-century chemistry in more than a thousand pages. Descartes's corpuscles, with their various figures and movements, could not inspire chemists with new questions or new techniques, only with *a posteriori* interpretations. Thirty years later, Wilhelm Homberg (who, unlike Lémery, knew that certain neutral salts, made with the help of an acid, could decompose and reconstitute an acid) limited himself to "pacifying" the imagery: the points of the acid no longer broke but entered the alkali, from which they could also be withdrawn, as a sword is inserted in a sheath. The composition therefore "masked" the acidity but did not destroy it.

A "Catholic Matter"

Robert Boyle is traditionally depicted in chemistry manuals and even in some histories of chemistry as the inventor of the modern notion of the element: it is said that he defined the elements as those indestructible bodies that form compounds and into which compounds can be decomposed. This reference to Boyle's "revolutionary" definition has been cited by contemporary historians as an example of the historical error of abstracting from its context a statement that sounds "true," i.e., acceptable, today (Kuhn, 1952; Boas, 1958). A rather different reading is thus suggested for Boyle's famous definition (from *The Sceptical Chymist*): "I now mean by 'elements,' as those chymists that speak plainest do by their principles, certain primitive and simple bodies, or perfectly unmingled bodies; which not being made of any other bodies, or of one another, are the ingredients of which all those called perfectly mixt bodies are immediately compounded, and into which they are ultimately resolved." In fact Boyle was not then attempting to define a rational (i.e., *our*) definition of an element but to refine the traditional definition in order to question it. He wondered in particular whether each element entered into the composition of all bodies, but he did so to cast doubt on the very existence of these "elementary" bodies, as commonly defined! In other words, Boyle was not substituting a modern idea of the element for the Aristotelian definition; rather, he was questioning the *function* of the element in chemical practice—namely, the idea of finding unities beyond diversity, both principles of creation and principles of the intelligibility of this diversity.

Boyle made a good choice in naming his book *The Sceptical Chymist* (1661). Atomist theory led him to a general skepticism toward any theory of chemistry, whether Aristotelian or Paracelsian, but also toward Lémery's Cartesian imagery—in fact, toward any attempt to found a theory of matter and its transformations based on what the chemist could do or observe.

Let us accept the definition of an element as a body that is not separable by chemical procedures and that enters into the composition of other bodies (but not of all bodies). In that case, silver, gold, and the other metals should be accepted as so many distinct elements, but the elements are no longer the universal principles of explanation for all compounds. And if we thus give up the idea that an element must enter into the composition of all natural bodies, how many elements are there? And how do we know that gold and silver are primitive bodies? They seem practically unresolvable, but are they real principles?

For Boyle, the consequence of mechanist atomism was that all chemical bodies, whether we could resolve them or not, were produced by "different textures" of a "catholic or universal" matter. The traditional analogy between atoms and letters of the alphabet lost its significance: if it was used it was not as letters in a text, beyond which there was nothing, but of printed typographical characters, all constructed from a single material. All the perceptible qualities, all the properties that chemists studied, were reduced to a coalition of particles we cannot qualify through chemical means, to the configuration, texture, and cohesion of their different arrangements.

But if everything was related to texture, if the chemist could no longer assume that the elements that he was attempting to separate were indestructible, his procedures lost their clarity and status as solid ground for argumentation. Boyle was the first to criticize the "dry way" separation: he pointed out that when compounds were put into a hot fire, the substances obtained were not necessarily components of them but "creatures of the fire" that revealed nothing about the original compound; the products were what experimenters call "artifacts" today. Slow agitation and gentle heat could transform textures and therefore produce elements different from the residual bodies of other methods of decomposition. Boyle went back to the alchemical tradition in this invitation to slow processes. In fact, like Locke and Newton, he actively sought the secret of transmuting metals into gold (Dobbs, 1975).

More than breaking with the past, Boyle's skepticism negated all the arguments used in past controversies. Everything became possible, but nothing was necessary. Chemists' practices were contingent, relative to their techniques, and the distinctions they produced had no essential value. The characteristics of bodies were not individual attributes of substances but the result of methods of arrangement and structure that the chemist could or could not transform. In the same way, the difference between mixts and elements depended on the chemist, not on nature: there was no reason for nature to make mixts from chemically homogeneous substances, and, correlatively, nothing could limit chemical transformations, the changes from one texture to another.

Boyle's atomism therefore implied the impossibility of chemistry ever becoming a science with a theory on which to base and explain its practices. The only possible chemical theory would in effect be mechanist and its subject would have to be textures, not the qualitatively different properties of bodies. Arrangements of particles "without qualities" would be responsible for what we call qualities. In the meantime, the chemist must be satisfied with functional definitions and criteria of identification, relative to what Boyle was the first to call "chemical analysis"; he must work to acquire an increasingly precise practical knowledge. So Boyle invented the "flame test," making it possible to recognize a substance by the color of its flame when it is burned, and he showed that in some cases the differences between the colors of compounds did not count, because they were due to the presence of impurities: what chemists characterized, for example, as green, white, and blue vitriol, were distinguished by the presence of iron, zinc, and copper.

A "Matter of Fact"

What could the chemist do? What distinctions did his experimental procedures afford him? These operational questions determined the categories in a chemistry inspired by analysis, but they cut off any relationship between the "element," as a principle of intelligibility, and the element as a body that the chemist could not decompose. That is why Boyle's disciples and certain historians were able to consider Boyle the first to make a modern definition of the element. It was a "negative empirical" concept (Thackray, 1970, p. 168, after David Knight) that reflected the limits of the technique of analysis. The idea of a negative empirical concept entailed a new epistemological strategy, however. It was a new kind of argument that grounded proof no longer on reason but on sheer experimental practice. The difficulty was in knowing who would accept this kind of argument, for whom the limits of analysis would be considered the chemist's horizon.

Developing the answer to this question was another aspect of Boyle's work. Boyle, a founding member of the Royal Society created in 1662 by Charles II, not only participated in this new kind of scientific organization but also, as Shapin and Schaffer (1986) have shown, actively connected it with the question of scientific proof.

Boyle is well known for his work on the elasticity of air, the volume of air being inversely proportional to the pressure. But he had to confront the criticism of those who, like Hobbes, recognized this work to be an argument in favor of the existence of the void, which they rejected. To answer his critics Boyle did not undertake to change the intellectual axiom that made the void an impossibility; he did not try to construct a rational concept of the void that would make it valid for everyone. Boyle did not address anyone and everyone, as do authors of traditional treatises, in an attempt to base the contention that "the void exists" on a conceptual demonstration. He used the air pump to show not the existence of the void but the functional possibility of decreasing the pressure of air. This demonstration took place in the laboratory, the place where the "fact," what Boyle called the "matter of fact," could be constructed from scratch. But this proof chose its own audience, those whom the air pump would indeed force to recognize *as a matter of fact* what a world without air would be like. These were *gentlemen* who had access to a laboratory, reliable witnesses to the facts constructed by the air pump. They were "colleagues," who from now on would use the air pump in their own laboratories. And the fact that "the void can be produced" would be recognized throughout Europe as less and less expensive and more and more dependable air pumps were produced. The very idea of "progress" in the construction of the apparatus assumed and created the truth of what it was supposed to accomplish.

It is here that the history of the experimental sciences, of which chemistry would one day be the fairest flower, began. But the practical and social invention of the experimental sciences is still not the point of departure that finally had in itself the power to create a "properly scientific" history for chemistry. Chemistry did not have the equivalent of a "pump": its problem was not the establishment of a fact, refinable at will, but the exploration of a jungle of "facts," in which materials, doctrines, procedures, and interpretations were mixed up together. And chemists had to face not skeptics who wanted demonstrations but rivals who pretended to understand, and who judged.

The End of Origins

Exploring a notion as rich in meaning as that of origins is always risky. In the case of chemistry this risk imposed itself because the question of the "alchemical origins" of chemistry haunts the memory of chemists and of our culture as a whole, as Lavoisier's "chemical revolution" haunts them. Everyone agrees that alchemy was not a "true science" but that Lavoisier's chemistry was. The "birth of chemistry," or chemistry's achievement of the title of science, should therefore take place somewhere between these two events, according to criteria that are the object of much controversy (see especially Holmes, 1989).

Our narrative rests on a definite choice, which is obviously not supposed to be the "true" choice. This choice depends on certain concerns that we would like to specify before setting, in a completely symbolic way, a date to mark the "end of the origins" of chemistry.

Balance Sheet

First we needed to bring alive the multiplicity of inheritances that make up chemistry. It is astonishing that techniques as dissimilar as metallurgy, pharmacy, and the art of the perfumer were ever assembled under a common denominator and melded into the common territory of "chemistry," with its own culture, methods, and identity. *A posteriori* the identity of a science always seems the most natural thing in the world. As we will soon show, however, chemistry's domain, the autonomy of its practices, and its identity in relationship to the other sciences have never ceased to be the object of historical struggles. At the "origin" of this problematic definition of chemistry there was already a long, involved, heterogeneous history—the history of alchemical practices. The polymorphous character of this history refutes the simplistic vision of alchemy being crushed by rationality when modern science was born.

Second, it was necessary to mark the continuities and breaks with the past carefully. The idea of the atom, as modern as it may seem to us, did not interrupt the continuity of chemical practice. Mechanical science did not whisk away alchemical traditions with a magic wand. They were not dead; they had been displaced or reformulated. Two "novelties" do mark the seventeenth century. On one hand, with van Helmont, John Mayow (1641–1679), Boyle, and then Stephen Hales (1677–1761), an interest in gases developed. At a time when quarrels over the void were all the rage, attention first centered on the vacuum pump, which gave researchers access to general properties such as elasticity. At the same time, new instruments entered the chemist's laboratory with gas—air pumps and bell jars for collecting the gases produced by chemical reactions—and they opened a new area for dispute. Besides, with the idea of the atom a new practical relationship to experimentation emerged, which meant that the interests of artisans and theoretical research diverged more and more: chemical transformations may produce demonstrative facts when they offer the possibility of getting the original product back, and the identification of such reactions became a common practice for the researcher/scientist while it held no interest for artisans.

Eventually, it became important to make a clear distinction between the use of reason, in the modern sense, and the status of science. For historians who treated the two themes as aspects of the same turning point—the "scientific revolution" responsible for the break with the past and the resemblance to our present—chemistry was a source of perplexity. As we will see in the following chapter, the figure of Lavoisier was suggested to be analogous to that of Galileo or Newton, but he arrived quite late, only at the end of the eighteenth century. In a work that was a pedagogical reference for a long time, *The Origins of Modern Science, 1300–1800* (1965), Herbert Butterfield stressed the difficulty by titling his chapter on chemistry "The Postponed Revolution in Chemistry." Just as it was necessary to reconsider the question of origins to appreciate the multiplicity of its inheritances, so it was necessary to separate the problem of identifying those who might be considered the first chemists (those who would claim to be practicing a rational science) from the problem of identifying a founding act that would have given chemistry the status of the progressive, cumulative undertaking we usually associate with a "true" science.

We have encountered no such founding act thus far, but the claims of rationality and of public knowledge abound, since it was transmission of knowledge, either public or esoteric, that led to a distinction between chemistry and alchemy. In the seventeenth century, where we will close this chapter, the question of the possibility of transmutation was not yet settled. At The Hague in 1667, Spinoza and Helvetius took the claims of a self-proclaimed gold maker very seriously. On the other hand, transmutation was no longer the central question for chemists, because they had much to consider in other sources, such as Agricola, Libavius, Lémery, and soon Boerhaave, for whom a "rational" account of the accumulated knowledge was the crux of the matter. In other words, the alchemical tradition, which was focused on deciphering the past, was not refuted but only marginalized. In the eighteenth and nineteenth centuries this social process of marginalization would turn into a way of defining chemistry: alchemy became "the other." A critical reference to alchemy became a commonplace that guaranteed the rationality of chemistry. In his *Dictionnaire de chymie* Pierre Joseph Macquer (1718–1784) wrote that "chemistry . . . happily has nothing in common but the name with the old chemistry, which is quite bad enough, for the same reason that it is bad for a girl who is full of spirit and good sense but very little known to have the name of a mother famous for her ineptitude and foolishness" (from the entry "Chymie," 1778 edition).

Options

We have pointed out that the rationalization of chemistry and the marginalization of alchemy are inseparable from the invention of printing. At the end of the seventeenth century, chemistry, having become "rational," had its "modern" writers and its public. Its courses were widely taught, its language was freed of all suspect references to occultism or the secrets lost or transmitted in cryptic terminology since the dawn of time. On the other hand, a new problem was developing, the problem of the status of chemistry. In effect this problem was inseparable from the opening, toward the end of the seventeenth century, of institutions in which the idea of science as a collective enterprise among recognized, credible colleagues, each judging the others, was developed. Printing had created the category of the public, and the new institutions in different ways created the public as spectators and chose those who would be the "actors," the scientists who not only would be capable of deciphering the "book of nature" but would be paid to do it and would have to persuade learned colleagues of the value of their discoveries. In the context of the academic institution, "rational" chemists would be confronted with the question of their place, of the identity that would be granted to their work.

Nicolas Lémery had been an apothecary and doctor. In 1699 he was named a member of the Royal Academy of Sciences. But in that same year the permanent secretary of the Academy, Fontenelle, wrote: "Chemistry resolves bodies into certain broad principles through visible operations . . . physics acts on those principles as chemistry has done on the bodies by delicate speculation; it resolves them into other, simpler principles . . . The spirit of chemistry is more confused, more hidden; it is more like mixts, in which the principles are more intermingled with each other; the spirit of physics is simpler and more acute, it goes back to the earliest origins, the other does not reach the ultimate."

With this double event in 1699—the entry into the Academy of Science of the man who incarnated the "rationality" of chemistry in France, and Fontenelle's subordination of chemistry to physics—we shall close our narrative of the origins of chemistry. This story ends with both institutional recognition and a judgment of chemistry's inferiority. But this judgment is formulated in terms that involuntarily admit the specific identity of what it condemns.

Apparently the Cartesian Fontenelle said the same thing as Boyle, the "sceptical chymist": the only real chemical science is that which goes all the way back to atoms with only geometric properties. Rather than contrasting the chemical process with atoms, however, his statement contrasts the two sciences of chemistry and physics. It advocates not the rationality of skepticism but a hierarchy of sciences that derives from the atomic hypothesis: chemistry could only be skeptical, while physics was assured of its principles since atoms were its territory. Fontenelle sends Boyle back to the "story of origins," substituting the reality of the rivalry between holders of scientific territories and titles for his fraternal utopia of gentlemen experimenters.

Fontenelle, however, described the subordination of chemistry using the vocabulary of chemical and even alchemical operations: spirit and principles. The existence of a specific cultural identity for chemistry was affirmed in the very words that attempted to deny it. The question that opened the eighteenth century was whether this cultural identity could become a scientific identity.

What would the French chemists who were members of the Academy say to their mechanist colleagues? How would they get their knowledge valued and their identity recognized in the face of the triumphant model of physics? How would they become a complete, separate science? How would they cope with an origin deprived of the halo of a Descartes, a Galileo, or a Newton? These questions were not raised by historians of chemistry but by the chemists themselves, and they never stopped asking them throughout the course of the eighteenth century.

2

THE CONQUEST OF A TERRITORY

Revolution!

At the end of the seventeenth century chemistry seemed to be a discipline precisely in the sense of "taught material," and this material was, for the most part, closely interdependent with medicine and artisanal practices such as metallurgy, perfumery, and the like. By the end of the eighteenth century chemistry was recognized as a separate science, autonomous, legitimate, based on a solid foundation and the source of applications useful to the public welfare. How did it acquire this status?

The quick and easy answer is: by a revolution. For two centuries chemists and historians have been arguing about the exact nature of this revolution and the name of its founder. Was it Stahl? Lavoisier? The group of chemists who developed pneumatic chemistry between 1750 and 1780? The disputes have been erudite and passionate, sometimes freighted with nationalistic fervor.

A New Paracelsus?

These disputes were all the more unresolvable because they concerned figures who had asked the same question themselves. From the beginning of his career, Lavoisier was convinced that his research should "bring about a revolution in physics and chemistry."[1] Twenty years before Lavoisier made this note, Gabriel François Venel stated, in the article "Chymie" in the *Encyclopédie* published in 1753, that chemistry was waiting for the "new Paracelsus," the bold, skillful, and enthusiastic chemist who would bring about a "revolution that would put chemistry in the place it deserves to occupy." Did Venel "predict" Lavoisier? In a sense he did—he certainly predicted the debates that would be set off by "Lavoisier's revolution," because he specified in a very pragmatic way the conditions that the so-called revolutionary would have to satisfy: "finding himself in a favorable position and profiting skillfully from fortunate circumstances, he would know how to get the attention of scholars, first by a clamorous show and a determined and positive tone, and then by the power of his reasoning, if his first weapons had breached their prejudices."

It was a "media" revolution that Venel was wishing for, not the austere, conceptual revolution that we like to imagine as the origin of a "real science." Lavoisier read the article "Chymie," in the *Encyclopédie,* as did all French chemists in the second half of the eighteenth century. It was a paradoxical situation: before beginning work, Lavoisier had read the essential gist of the accusations that would be brought against him by those contemporaries and historians who refused to grant him the title of a "true" revolutionary. As a "media revolutionary," he indeed knew how to turn a fortuitous opportunity into "reason" (i.e., a new conceptual foundation for chemistry); he knew how to use the controversy that he raised over the doctrine of "phlogiston" as the starting point of a revolutionary, loud, and ostentatious claim to a new beginning for chemistry.

The conceptual history of science is generally viewed as the history of a "process without a subject," as Althusser said. The subject, with his intentions, his psychology, and his illusions, is on the side of opinion and error. The truth of the concept means that it defines the subject, thus reducing the biography of the particular person who advanced it to an insignificant anecdote. It was in this light that Lavoisier intended to present the situation: on one side, tradition, weighed down by illusions and subjectivity; on the other, the asceticism of a scientist who made a clean break with tradition and became the anonymous spokesman for nature—the scientist who was, in other words, the first to put himself into a condition to hear her. Lavoisier was nothing; the balance did everything, and legitimized Lavoisier's assertions. Again the situation was paradoxical: did Lavoisier illustrate Bachelard and Althusser, or did they use the scenario constructed by Lavoisier to ratify the break between him and his predecessors? The beauty and the ambiguity of the history of science is that its subject is the activity of "those who produce history," of people who ask themselves the questions asked by historians, people who often go so far as to integrate a "theory of history" into their rhetorical-scientific strategies.[2]

Lavoisier's operation was remarkably effective. Traditional descriptions of eighteenth-century chemistry are focused on the struggles between "them" and "us": between those who believed in an erroneous doctrine, the theory of phlogiston, and those (ourselves) who since Lavoisier have known that that theory was false. In brief, eighteenth-century chemistry was most often described as "waiting for Lavoisier."[3]

A Useful Break

The temptation to read eighteenth-century chemistry backward, starting with Lavoisier, is reinforced by the difficulty of reading texts written previous to his work (Crosland, 1962). One becomes lost, without landmarks, in a jungle of exotic and obscure terminology. Some products were named after their inventor (Glauber's salt, Libavius's liquor),[4] others after their origin (Roman vitriol, Hungarian vitriol), others after their medicinal effects, and yet others after their method of preparation (flower of sulfur, obtained by condensing vapor, precipitate of sulfur, spirit or oil of vitriol, depending on whether the acid was more or less concentrated). On every page one wonders: what is he talking about? What substance is he describing? We ask, in effect, what does it correspond to in the modern nomenclature created by Lavoisier and his colleagues Berthollet, Louis-Bernard Guyton de Morveau, and Antoine Fourcroy in 1787? Even if one manages, after considerable apprenticeship, to translate and decipher these texts, the nomenclature of the resultant products is still a headache; in fact, it was already denounced as such throughout the eighteenth century.

The defects are obvious to us: several names for what we know to be only one product, several substances with the same name. In fact these problems could only begin to appear as chemists' practices changed during the eighteenth century. The jungle of traditional names was no denser than the jungle of techniques, because the techniques were determined by the problem of extraction. There was no chemical supply store where one could buy nitric acid or calcium carbonate. The chemist learned to extract products from primary materials, and each name designated a particular recipe for extraction, a recipe that had to be followed step by step, for each recipe began with a particular primary material and most often resulted not in a "pure body" but in a mixture that was almost as individual. Besides, the same name could indicate different products capable of playing the same role in the same class of preparations. Redundancy and homonyms became apparent and were criticized as such when individual and nonreproducible techniques for extraction and the treatment of primary materials gave way to laboratory methods of identification. Then the identity of a product no longer depended on its method of preparation or its phenomenological manifestations—whether flower or precipitate.

Differentiating between the properties of a body and those that it gets from "impurities," making exhaustive inventories, classifying substances, reactions, colors, procedures—these tasks undertaken by eighteenth-century chemists made the "rationalization" of chemical nomenclature a necessary enterprise, which had been under way for more than forty years before Lavoisier's reforms. Guillaume-François Rouelle, Pierre Joseph Macquer, William Cullen, Torbern Bergman, and others endeavored to refine traditional nomenclature by eliminating redundancies and introducing generic names, like "vitriols," for example. The first nomenclature reform answered a need that had been expressed for a long time and fit into a series of attempts made all over Europe.

To avoid wandering through the eighteenth century backward while waiting for Lavoisier to arrive on the scene, we must get away from the traditional scenarios centered exclusively on phlogiston and nomenclature; we must give up the image of Lavoisier blowing up the dam that tradition had built against progress. The fact that this is still a challenge today is the best proof of Lavoisier's success. He managed to limit his successors and most historians of chemistry to a narrative that is punctuated by a "before Lavoisier" and an "after Lavoisier." But having offered this homage, one can and must ask oneself about the many reasons for this success and analyze what constitutes the "before" and the "after."

The first question, which relates to the questions posed at the end of the previous chapter, is: how did chemists meet the challenge of mechanist atomism and proclaim the autonomy of their science before Lavoisier?

"Query 31"

In 1704, Isaac Newton, who was then president—and "dictator for life"—of the Royal Society of London, published *Opticks*, a treatise on the "Reflexions, Refractions, Inflexions and Colours of Light." This book (or, rather, the enlarged edition published in 1717) would become the primary reference for experimental physics in the eighteenth century. The work was not concerned with the mathematics of motion but the study of electrical, magnetic, chemical, and even biological and geological phenomena. With the *Principia* in 1687, Newton had dared to introduce unintelligible forces acting at a distance in the rational universe of "the motions and figures of heavenly bodies"; in the *Opticks*, he gave his seal of approval to an experimental approach that, far from considering nonmechanical phenomena the "ambiguous and shrouded" expression of an underlying mechanical reality, in the manner of Fontenelle, assumed them capable of "proving" hypotheses about themselves.

At the end of the *Opticks* Newton tackled a certain number of controversial subjects in the form of "queries," notably, in the famous Query 31, chemistry.[5] Theoretical interrogation and experimental practice were closely linked in it:

> For when Salt of Tartar runs *per Deliquium,* is not this done by an Attraction between the Particles of the Salt of Tartar, and the Particles of the Water which float in the Air in the form of Vapours? And why does not common Salt, or Salt-petre, or Vitriol, run *per Deliquium,* but for want of such an Attraction? Or why does not Salt of Tartar draw more Water out of the Air than in a certain Proportion to its quantity, but for want of an attractive Force after it is satiated with Water? And whence is it but from this attractive Power that Water which alone distils with a gentle luke-warm Heat, will not distil from Salt of Tartar without a great Heat? And is it not from the like attractive Power between the Particles of Oil of Vitriol and the Particles of Water, that Oil of Vitriol draws to it a good quantity of Water out of the Air, and after it is satiated draws no more, and in Distillation lets go the Water very difficultly? (Newton, 1952, pp. 376–377)

The hierarchy based on the mechanical atom without qualities was therefore abandoned. Newton accepted the "solid, heavy, hard, impenetrable, mobile" particles that God very probably made at the beginning, but he did not accept the idea that they were characterized only by the "force of inertia," "a passive principle in virtue of which bodies remain in movement or at rest." In that case there would be "neither destruction nor generation, neither vegetation nor life." The accelerating motion of the celestial bodies, like chemical phenomena and light, escapes from mechanics and arises from "active principles."

Two Readings

"Query 31" can be seen as an extrapolation of gravitational physics to chemistry. It would be read that way in England. The English disciples of Newton, like John Theophilus Desaguliers, who was the Curator of Experiments of the Royal Society of London during the last years of Newton's life, would try to explain chemical and later electrical and magnetic phenomena in terms of forces—which meant that those aspects of chemical phenomena that illustrated the working of an attractive or repulsive force in a clear way would be favored.

Another reading is possible, however: Newton gave back to chemists the right to speak of power, of the force of reagents—in other words, to make sense of their practices, of their methods, which was forbidden by purely mechanistic chemistry. In fact, while gravitational attraction's primary attribute is uniformity (the same force comes into play between the apple and the earth, the moon and the earth, the earth and the sun), this uniformity disappears in "Query 31" in favor of a strictly chemical measurement of forces: chemical reactions make it possible to compare the forces that effectively unite the particles of a compound; it is the reactions, and not theoretical calculations, that are the basis of comparison. In the same way, the variety of forces makes it possible to understand the possibilities and the limits of chemical analysis:

Now the smallest Particles of Matter may cohere by the strongest Attractions, and compose bigger Particles of weaker Virtue; and many of these may cohere and compose bigger Particles whose Virtue is still weaker, and so on for divers Successions, until the Progression end in the biggest Particles on which the Operation in Chymistry, and the Colours of natural Bodies depend, and which by cohering compose Bodies of a sensible Magnitude. (Newton, 1952, p. 394)

Moreover, Newton introduced in the same query the hypothesis of a repulsive force, the existence of which is attested by both optical and physicochemical phenomena: thus, the particles of fixed bodies are kept together by attraction, but the "prodigious expansion" of the particles in volatile bodies, the fact that once the sphere of attraction is broken, they spread out rapidly, testifies to the existence of a repulsive force that "begins to act where the attractive force has just ceased." Fermentation, the action of heat, and even the fact that flies walk on water "without getting their feet wet" are phenomena recognized as authorizing, as well as the mathematics of planetary movement, a new type of causality. Newton took care to distinguish his behavior from that of the "Aristotelians" who spoke of occult qualities:

> To tell us that every Species of Things is endow'd with an occult specifick Quality by which it acts and produces manifest Effects, is to tell us nothing: But to derive two or three general Principles of Motion from Phaenomena, and afterwards to tell us how the Properties and Actions of all corporeal Things follow from those manifest Principles, would be a very great step in Philosophy, though the Causes of those Principles were not yet discover'd. (Newton, 1952, pp. 401–402)

Contrary to the occult qualities that explained each particular property by a particular virtue, active principles unify a number of apparently disparate phenomena. The two divergent readings of "Query 31" agree on this difference but not on its interpretation. Newton's English disciples kept to the idea of mutually attractive or repulsive forces. But Venel wrote in the *Encyclopédie* that when chemists claim the right to "make a cause" from a "certain number of relative effects of the same order," they do the same as Newton (and Aristotle). In France the *Opticks* did not simply put a seal of approval on experimental investigation. Throughout the course of the eighteenth century it nourished protests against the "establishment," which, according to Venel, did not know what Newton knew: "that nature achieves most of her effects by unknown means; we cannot count her resources; and the only real absurdity would be to wish to limit her, by reducing her to a certain number of principles of action and ways of operating." The irony was that "Query 31," the starting point for diverging histories in the eighteenth century, was itself the end of a secret history centered on the tradition that would be condemned by the eighteenth century: alchemy.

Newton's Secret

Was Newton an alchemist? The word makes some people fear—not without reason—that this historic fact will become a shibboleth that will confuse the boundaries between scientific rationality and irrationality.[6]

Newton's historians have always known that he had a weakness for alchemy but had treated the subject with the greatest discretion. The discretion ended in 1946, during Newton's tricentennial celebrations. John Maynard Keynes, who had bought a large number of Newton's alchemical manuscripts in 1936, when they were about to be dispersed in a public sale, declared that "Newton was not the first of the age of reason. He was the last of the magicians, the last of the Babylonians and Sumerians, the last great mind which looked out on the visible and intellectual world with the same eyes as those who began to build our intellectual inheritance rather less than 10,000 years ago" (cited in Dobbs, 1975, p. 13).

Since then historical studies have multiplied. In 1958 Marie Boas could still pretend that Newton was not an alchemist: he was simply interested in the chemistry of metals. And certainly Newton's experiments have a meticulous, quantitative precision unsullied by triumphal declarations, descriptions of prodigious transformations, or manifestations of a credulous enthusiasm. He was an austere researcher who sought to pierce the enigma of the activity of matter, not a crank mistaking his desires for reality. It is, however, also true that Newton eagerly attempted to decipher, annotate, and understand the most esoteric and enigmatic manuscripts, not those in which we can recognize a "modern" outlook. (William Newman [1994] has recently described the many inspirations Newton found in the works of Eirenaeus Philalethes, which in fact were written by George Starkey, a colonial American physician who spent most of his adult years in England and who died in London's Great Plague of 1665.) He must therefore have been convinced that the alchemists *had* a secret and, what is more, he worked on the most impenetrable secret of all himself.

Richard Westfall was the first to dare ask the question that will always remain without a definitive answer but has spurred Newtonian studies nevertheless: what if the *Principia* was not for Newton himself his career's climax but more like an interruption of his primary labor? What if he did turn to celestial motion as to a simple case of the "active principles" of matter, whose secrets he was tracking in his Cambridge laboratory? We cannot analyze Newton's alchemical work here. Betty Jo Dobbs has shown that it centered on the "green lion" (a symbol for the star regulus of antimony, or metallic antimony), made famous in *The Triumphant Chariot of Antimony* by the fifteenth-century alchemist Basil Valentine. Newton would have seen the star as the sign of a power of attraction and hoped that the regulus, drawing from the aerial "universal Spirit" the "fermental virtue" of its "spiritual semen," could purify the common mercury and turn it into "philosophical mercury." Newton's alchemy implied the notion of an attractive power and also the idea of a "mediating" agent, Diana's doves (the purest silver), without which common mercury and metallic antimony could not be "sociable." Unlike the "forces" that we know, however, the alchemical "mediator" could have an existence apart from the bodies it animated.

For those who are interested in astronomy and the science of motion, the figure of Newton as alchemist is anecdotal. But for the history of chemistry it is central. On the one hand it explains the fortunate "repercussions" of Newtonian physics that gave chemists what they needed so much: an understanding of their operations in terms of the power of reagents. On the other hand, it illuminates the choice that would face eighteenth-century chemistry: whether to submit itself to forces of the Newtonian type or to demand from chemical phenomena the "principles" that make them intelligible. Newton himself must have hesitated. While he was searching the sky to learn how to describe the action of specific "forces," he arrived at the astonishing realization that a single, uniform force was sufficient to explain the observed motions; this force therefore defined the bodies it acted upon as homogeneous and uniform, like the mechanical atoms. "Query 31" can thus be read as a precarious compromise between the necessity of relevance for the alchemist/chemist at his furnace and the temptation to extrapolate to earthly phenomena the principle of order that had, in an unexpected way, triumphed in the sky.

The Salts: Relationships and Displacements

In the beginning of his *Considérations générales sur la nature des acides* in 1777, Lavoisier presented a historical survey and acknowledged an achievement of the past that he would never wipe out: the theory of salts.

"This theory," he wrote, "is so well established today that it can be seen as the most certain and complete part of chemistry" (Lavoisier, 1864, vol. 2, p. 248). He himself intended to do for the constituent principles of neutral salts what his predecessors had done for the salts themselves.

This reverence for the past is all the more remarkable in that the seventeenth-century notion of "salt" appears extremely vague and obscure to a modern chemist. Salts were characterized first of all by their solubility in water. What is for us "acid," "spirit of salt," or "spirit of vitriol" was then taken as salt; on the other hand, many substances that to us are salts (for example, carbonates) were then considered "earths," because they are insoluble.

Academic Chemists

What is a salt? French chemists did not try to answer this question directly. No doubt it would have led them to Boyle's skepticism about principles, but perhaps not to new practices. On the other hand, beyond the Cartesian imagery of broken points or swords being sheathed, the question asked by the corpuscularians—what has become of the properties of the components when the compound does not exhibit them?— gave them a new interest in a particular category of salts: those that are neutral (formed from an alkaline salt and an acid salt).[7] Willhelm Homberg, the most productive of the academic chemists of the period, called these compounds "middle salts."[8]

Leaving aside "speculative" questions to immerse himself in the chemistry of salts was, according to Holmes (1989, p. 33–55), the answer to the new challenge that chemistry was facing "in order to become a science," to be legitimized as one of the disciplines of the Academy. Academics could no longer be content to produce didactic works and put established facts in a rational order. They had to produce original results. In 1699 the new rules of the Academy required them to contribute directly to the advancement of the sciences through their work and to communicate new results regularly to the Academy. The publication of an annual volume of the *Histoire et mémoires de l'Académie* began in 1700.[9] The study of salts would be a popular topic, because of the promise of new products.

The middle salts subverted the idea of the salt as a principle. Homberg asserted that they could result as well from the combination of salts (spirit of vitriol and tartar salt, for example) as from an acid and an alkaline earth or an acid and a metal![10] In other words, the study of middle salts led to the idea of the interchangeability of bodies corresponding to distinct categories of principles. It was an experimental field in which the possibility of establishing relationships could replace the "principle" as the standard of "explanans."

The Table of Relationships

In 1718 Etienne-François Geoffroy, member of the Royal Academy of Sciences and of the Royal Society, presented to the Academy a *Table des différents rapports observés entre différentes substances*. The reception was cool. In his *Eloge de Geoffroy* in 1731, Fontenelle emphasized that these relationships "upset some people who feared that these were disguised attractions, all the more dangerous because clever people had already given them seductive forms." It was a legitimate fear. In 1706–1707 Geoffroy had in effect lectured on the *Opticks* in ten sessions before the Academy. And his table followed the model proposed by Newton in "Query 31": at the head of each column a body was named, followed by all those that could be combined with it. The order was determined by what Newton had called their respective "attraction" for the body at the head of the column: in combining with the head body, a body "displaced" all those that followed it and "was displaced by" those that preceded it.

The idea of "displacement" transcended the strict lessons of experiment, from which Geoffroy's table was officially derived. Speaking of displacement meant considering the chemical substance as a composition-combination, and the reaction in terms of association and dissociation. Since Aristotle's time the properties of a body had been rooted in a "principle," a substance that acted and explained; now everything took place on the level of relationships. The traditional notion of "affinity" was correspondingly transformed. Far from deriving from principles, Geoffroy's affinity-relationship called into question principles in their role as agents explaining operations. The relationship designated the combination as a principle of intelligibility: combination as a state of union between two bodies would result from combination as a process of union or disunion. The process made it possible to characterize the state according to a scale: the affinity or relationship was "stronger" in one combination than in another. The "matter" had to be understood on the basis of the relationship and could be explored according to the interplay of the relationships, creating or destroying combinations.

Geoffroy's table, if it was not a simple derivative of empirical experience, a simple putting in order of known reactions, nevertheless corresponded to the empiricist reading of "Query 31." Contrary to the English Newtonians, Geoffroy did not seek proof for the existence of forces in chemistry. He limited himself to envisaging chemical reactions from a point of view that displaced the explanatory function from principles, which were responsible for qualities, in favor of relationships. The fact that Newtonian attraction was not well considered in corpuscularian France is not perhaps the only explanation for this strategy. Most of the reactions put in order in Geoffroy's table produced the "neutral" or middle salts that he and his colleagues had been studying since Homberg. As we have pointed out, those reactions defined bodies that answered to traditionally distinct categories as interchangeable. And that is what Geoffroy's table presented; for example, it put metals, absorbent earths, alkali salts, and mineral sulfur in the sea salt acid column, all considered from the same point of view. In other words, Geoffroy's relationship was not just a disguised Newtonian attraction. It was an attraction redefined within a preexisting, significant context, that of the chemistry of salts.

The chemistry of salts, like the table of relationships or affinities that would henceforth accompany it, marks therefore the inception of a new *systematic practice*. It was not a question of a "system" in the sense of an explicit presentation of definitions and conceptual choices. But neither was it an exercise in empiricism, in the sense that the results would be independent of any conceptual scenario. The corpuscularian question— how were acid and alkaline qualities mutually neutralized? (a question that had drawn attention to neutral salts)—would be abandoned rather quickly. The neutral or middle salt, product of a displacement reaction, *organized* by its functional definition a true research program that transformed the practices of chemists and their criteria of intelligibility.

Salts Redefined

In 1736 Henri Louis Duhamel succeeded in isolating the free (nonvolatile) alkali that entered into the composition of sea salt. He showed that this alkali formed marine salt, but also other salts that could be distinguished from analogous salts made from alkali derived from tartar salt (in modern terminology, Duhamel distinguished between salts formed from sodium and potassium, respectively). This demonstration involves a judicious series of displacements, one salt producing another salt. It enriched the table of relationships but also gave a new operational meaning to the "re-creation of a body in its primitive state," which had been used as an argument for corpuscular chemistry. The alkali came from marine salt and, when combined with marine acid, it produced a salt that could not be distinguished from marine salt. A body obtained by extraction could therefore be re-created in the laboratory. The "primitive state" no longer corresponded here to the cyclic idea of a regeneration that would prove the indestructibility of the components; it now referred to the practical production of new criteria for identifying a body. Marine salt was no longer from the sea except by usage; it could be produced in the laboratory by means other than extraction. The chemist could henceforth make a distinction between the intrinsic properties of a body (independent of the means of production) and those that resulted from the source from which it was extracted (the impurities). Boyle's skepticism linked the identification of bodies with the limits of our powers of analysis and synthesis. Analysis and synthesis now had a demonstrative value: establishing the identity of bodies and suggesting a "rational" way of renaming them. For example, on the basis of Duhamel's work, Macquer distinguished between marine salt and "common salt from vegetable alkali."[11]

Correlatively, the "salt" lost its definition as a principle based on its solubility in water: even insoluble products, when obtained from a displacement that identified them as analogous to known middle salts, would be recognized as salts. Solubility thus became a way of separating salts. The difference between alkaline earths and tartar salt, united by displacement operations, was abandoned as well.

Now we understand better why Lavoisier wanted to do for the "saline principles" what had already been done for the neutral salts. But we must beware of this too easy, too direct understanding. Let us not forget that Lavoisier himself chose the chemistry of salts as a focus of his work among other aspects that we have not discussed. In particular, we have just condemned with a stroke of the pen the explanatory value of the "principles," although many chemists who used Geoffroy's table in the eighteenth century did not abandon this notion. For instance, in Macquer's *Dictionnaire de chymie*, the principles kept their essential functions: they retained their integrity through different compositions and decompositions and conveyed the properties of bodies. They did not explain affinity but were characterized by the affinities that they had for each other. "The affinities of the four principles forming two new compounds can, by a mutual exchange, cause two decompositions, two new combinations. That happens every time the sum of the affinities that each of the principles of the two compounds shares with the principles of the others is greater than that of the affinities that have among them the principles that form the first two compounds" (cited by Anderson [1984, p. 58], whose analysis we follow here). Macquer certainly submitted the principles to the "quantitative" problematics of affinities, but he regarded the affinities as purely empirical. Not only were Geoffroy's affinity-relationships distinct from disguised attractions, but they did not yet have the power to redefine French chemical doctrines. At the time when Macquer was writing his dictionary, Newton was considered a mechanist, along with the corpuscularians; and the forces of attraction, like the little bodies defined by their motions and figures, were "only mechanical" agents. French chemists defined themselves by their relationship to the crux of Stahl's chemistry.

Principles: Elements and Instruments

Georg Ernst Stahl (1660–1734) was both physician to the king of Prussia and a chemist. As a professor of medicine at the University of Halle, he wrote systematic treatises for medical practitioners and presented his observations and works in the hope of improving practices. He was more interested in procedures than in proofs. He was therefore a stranger to the new academic customs that were being instituted in France and England. But he left his mark on the history of medicine as much as on that of chemistry, and in both cases that mark had primarily a polemical dimension. He revitalized traditional arguments, using them as weapons against mechanism's claims to explanation. Thus Stahl is often considered the father of the "vitalist" theory, which opposed the comparison of living beings to machines that could be explained in physical and chemical terms. For him the only thing that physics and chemistry could explain was corruption—the decomposition of a living body after its death. As for life, it demanded its own principle, a particular cause that fought spontaneous corruption. Stahl added to and systematized the doctrine that he had learned from his predecessors Kunckel, Johann Glauber, and especially Johann Becher. He mostly ignored the salt chemistry of the French corpuscularians and gave such terms as *mixt, principle,* and *affinity* a new, antimechanist relevance.

In the history of the life sciences, references to Stahl continued up to the twentieth century. The influence of his chemical doctrine was more limited in time but wider. Stahl was a source of inspiration and discussion during the second half of the eighteenth century. So when Lavoisier chose the "phlogiston" element-principle associated with Stahl's name as the target of his attacks, it was not a decaying relic of tradition that he challenged but a work that symbolized for most chemists the achievement of autonomy for their science.

Stahl's Principles

According to Stahl, the special characteristic of chemistry was the "mixtive union," or the mixt, which had to be distinguished from the aggregation. An aggregation was only a mechanical union. Whether it was understood in terms of interlacing corpuscles or Newtonian attraction, an aggregation was explained by the general properties of masses and movements—in other word, by mechanics, the science of essentially homogeneous materials. On the other hand, the mixt implied the qualitative diversity of substances that could only be analyzed by changing their properties. In effect, mixtive union created new homogeneous bodies from heterogeneous ones and could not be understood in terms of simple spatial proximity among the particles. The analysis of mixts and the characterization of their principles was the task of the chemist alone.

Following his master Becher, Stahl recognized two principles of all mixts: water and earth. But he distinguished three different types of earth (Metzger, 1930, pp. 130–138): vitrifiable earth, the principle of fusibility that went back to the heavy solidity of minerals; sulfureous or phlogistic earth, which was light and flammable; and mercurial or metallic earth, which gave metals their malleability and brightness. The identification of these principles was related to the ancient theory of affinity: if acids attacked metals, for example, it was because they were analogous with the metals, because they shared a principle with them. And it was therefore also appropriate to account for the properties of bodies in terms of the absolute qualities of principles.

According to his disciple Henckel, who shared his doubts, Stahl hesitated over the status of the third earth, the metallic one. This hesitation showed that the alchemical investigation focused on metals had been swept away by Stahl's great innovation: that the corrosion of metal and the combustion of wood or charcoal relate to the same phenomenon. The metal "burns" (slowly). Corrosion causes it to lose its light and volatile phlogiston, as it causes mineral coal to do the same. The only difference is that a return to the primitive state is possible, by reabsorption of phlogiston, in the case of the metal. But this return to the primitive state itself argues for the identity of combustion and corrosion. Charcoal, saturated with phlogiston, gives back the shine to metallic calx: it restores its lost phlogiston. Mining and metallurgical technologies for the "reduction of ores" could also be explained. When metals lose their phlogiston, they also lose their metallic aspect. Did this mean that metals consisted only of the first two earths? In that case, why all the compounds of these two earths are not metals remains to be explained. Is it a question of the proportion, or "concoction," of the phlogiston? In brief, Henckel concluded: "We are justified in wondering with Stahl if the third earth is generically or numerically different from the two others" (cited in Metzger, 1930, p. 133).

From the Chemical "Point of View"

In Stahl's chemistry an agent bridged the gap between aggregate and mixt. This was the "instrument." Fire (or heat), water (as a solvent), and air intervened as mechanical agents; they made the mixt possible but did not cause it. Fire put the phlogistic earth in motion, air blew off the most volatile parts of the bodies, water put the parts of the solution into motion. The idea of the body-instrument made it possible to give a limited place to the mechanist explanations in contrast to which Stahlian chemistry defined itself.

The distinction between aggregation and mixt was going to make a big difference. Its first effect was to put Newtonian attraction back among the mechanical agents that explained aggregation, to mix up in the same category Cartesian chemists like Lémery, mechanists like Boyle, or Newtonians like Boerhaave (the first chemist on the continent overtly to adopt Newton's position)[12] and John Theophilus Desaguliers, John Keill, and John Freind of England. To explain aggregations in terms of attractive and repulsive forces was still to explain them in terms of juxtaposition or dissociation—that is, to ignore the question of the chemical "bond" that created this or that body from other bodies. Venel's article "Chymie," in the *Encyclopédie,* would be based on this opposition between aggregation and mixtion. Correspondingly, the Newtonian roots of relationships or affinities shown in the tables would be forgotten. Mixtion between principles took place according to the laws of affinity, and these laws had links only with the empirical practice of chemists.

In 1758 the Academy of Rouen offered a prize for the answer to an apparently modest question that in fact involved all of chemistry: "Determine the affinities that are found among the major mixts as M. Geoffroy began to do, and find a physico-mechanical explanation for these affinities." The jury would have to recognize that no candidate had been able to satisfy both challenges and would name two winners, who approached the problem, respectively, from the point of view of a "physico-mechanical system" and from that of chemistry. For physics, Georges-Louis Lesage imagined an ingenious system in which affinity was explained by the different porosities of the bodies submitted to continual shocks of small particles, the "ultramondaine corpuscles" that also were supposed to explain attraction at a distance. For chemistry, Limbourg, from Liège, presented a new table of thirty-three columns. Limbourg rejected the mechanist interpretation of affinities *à la* Lémery and stressed the analogy with the Newtonian forces of attraction. Perhaps, he said, the same property is considered by physicists from the point of view of its effect on masses, and by chemists from the point of view of its effect on the elements. This "difference of points of view" meant that the Newtonian attraction could explain nothing as long as the difference between the physical mass and the chemical element or principle had not been elucidated. The chemical point of view therefore commanded attention, since it could claim an autonomy that Newtonian physicists could not hope to reduce.

If Stahl's polemic against the reduction of chemistry to physico-mechanical explanations met with complete success in France, the French identification of principles, on the other hand, although attached to his name, did not follow his doctrine. In the article in the *Encyclopédie* that he devoted to "Principles," Venel announced that chemists in general recognized four principles: earth, air, water, and fire, which they called "phlogiston," "with the Stahlians." But for Stahl and his disciples Henckel and Juncker, phlogiston was an earth and fire an instrument. What was called "Stahlian chemistry" in mid-eighteenth-century France certainly was inspired by Stahl's theory, but as it had been transformed by a man who was presented as his faithful interpreter, Guillaume-François Rouelle (Rappaport, 1961).

Phlogiston

Rouelle (1703–1770) was not an academician. He became aware of Stahl's work (written in a crude mixture of German and Latin) in an indirect way, in a publication by Senac in 1723. Like Lémery, Rouelle was originally an apothecary, and his reputation was made on the basis of the private course he gave before being named *demonstrateur* at the Jardin du Roi (Jardin des Plantes). His courses, like those of Lémery, attracted everyone who counted in Paris. Rousseau and Diderot learned chemistry there, as did Turgot, Macquer, Venel, and Lavoisier. In brief, a whole new generation accepted the academic work on the chemistry of salts as well established, purely empirical, and without great conceptual significance and discovered another universe through Rouelle,[13] that of the "Stahlian revolution."

Curiously, the only point of Rouelle's teaching—known through Diderot's notes—that stood up to the criticism of his own pupils, such as Venel and Macquer, was one that owed nothing to Stahl's doctrine: the association between phlogiston and fire. The system taught by Rouelle re-created the master's doctrine and rearticulated it according to a systematic association between two notions: the element-principle and the instrument. More precisely, Rouelle attributed two functions to principles: that of forming mixts and that of being the agent or instrument of chemical reactions.

Thus the four principles, earth, air, fire, and water, were principles both of the chemist's operations and of the mixts they operated upon. As instruments they were, unlike specific chemical reagents, "natural and general," always at work in every chemical operation. As constituent elements, they did not contradict the chemistry of displacement but transcended it: the chemist could never isolate or characterize an element as he characterized a body; an element was not isolable, for it could not be separated from a mixt without re-creating a new mixt in the process.

Fire or heat was therefore the instrument, and phlogiston was the element entering into the composition of mixts. Phlogiston, as an element, explained combustion, as well as the transformations of calx into metal and metal into calx.[14] That earth and water were elements went without saying. That air was an element was another innovation, in comparison with Stahl's ideas. By giving a chemical role to air, Rouelle took into account the work of Stephen Hales of England, who, in *Vegetable Staticks*,[15] studied the "air" given off by the fermentation of vegetable matter and by certain chemical reactions. Following Hales's example, Rouelle learned to collect airs and even improved the English apparatus. But if the role of elements was easily recognized for air, earth, water, and fire, the possibility of defining them all, in the same way, as instruments was more difficult. There was no problem for fire, for it was the most traditional, unanimously recognized instrument of the chemist.[16] *Ignis mutat res.* Water, earth, and air, in their role as *instruments,* were not at all clear. To illustrate the role of earth, Rouelle went so far as to point to the vessels, without which chemistry would not be possible.

Rouelle's theory was not, therefore, centered on phlogiston, any more than Stahl's. But phlogiston was still the "revolutionary" element, since it was associated with Stahl's unanimously admired discovery[17] that combustion and corrosion were identical, and his identification of the inverse operation (what we call "reduction") demonstrated the value of Stahlian chemistry, which refused to submit to the mechanical model. Moreover, the phlogistic fire, which was Rouelle's personal contribution, was the only case in which the element-instrument association was well balanced, without recourse to an argument whose artificiality would be denounced by Venel and Macquer.[18] So it was not the heritage of a secular tradition but the most representative product of the "chemistry of the Enlightenment" that Lavoisier would attack.

Although Rouelle's association of element and instrument was not new, it was representative of the values claimed by the French chemistry of the Enlightenment. In effect, it illustrated the double autonomy that this chemistry demanded from the mechanical model: the chemist "dirtied his hands," he operated at the level of mixts—he was nothing without his instruments, and he had the grace to acknowledge it; the chemist was interested in the intimacy of materials, in the elements, which could not be isolated because they belonged to the order of quality, of production, of the heterogeneous. In this sense, the chemistry of the Enlightenment disseminated by Rouelle was the product of a double polemic, that already undertaken by Stahl as a response to mechanism and that against "academic," elitist, and abstract science.

By Chemistry Possessed

The tensions between laboratory chemistry and the ideal of an academic career are not new. When Hermann Boerhaave exclaimed during his inaugural address at the University of Leiden in 1718, "I am going to speak of chemistry, of chemistry!" (cited in Meinel, 1983), he was apologizing for introducing—to a congregation of the most scholarly professors, practitioners of the most perfect sciences—a science that was crude, coarse, and difficult, far removed from the purview of scholars. Chemistry was unknown or suspect in the eyes of these scholars, who disdained its preoccupation with fires, smoke, and cinders.

This exclamation, like the apologies, was somewhat rhetorical. Boerhaave was a renowned teacher, and his *Elementa Chemiae* (1724) was foremost among the books from which eighteenth-century chemists learned their trade.[19] But the same theme reappeared throughout the eighteenth century: chemistry was work, in the sense in which "work" at that time meant the kind of painful and slavish labor that humbled those who practiced it.

The Value of Chemistry

This theme—chemistry as *work*—was part of Enlightenment chemistry as well, but its rhetorical sense was transformed: its practitioners thought of themselves not as lower-level scholars but as champions of the uniqueness of chemistry. Significantly, in his 1785 eulogy of Macquer, Vicq d'Azyr used Boerhaave's rhetoric, but in reverse, against professors who found it beneath their dignity to subject themselves to the chemist's labor: "When the progress of knowledge forced them to leave the school to question nature in laboratories, they thought it proper to their status to appear in their scholarly gowns; by wearing this garb they reduced

themselves to the impossibility of doing anything but speechifying"
(cited in Meinel, 1983, p. 125).

This rhetorical transformation was part of an offensive that was epistemological, philosophical, political, and social all at once, an offensive
against the model of the "professors' science," the science of those who,
as Diderot wrote in *De l'interprétation de la nature,* "reflect," "have many
ideas and no instruments," and look down on those who "bustle about"
and have "many instruments and few ideas."

What is the "value" of a science? Does the value reside in the certainty
of its principles, in the power of *a priori* judgments regardless of the need
for "bustling about"? Chemistry was confronted with this problem. It
accompanied the prodigious development of mining activity in Sweden
and Germany. In these countries chemistry laboratories and assay and
testing offices multiplied, and numerous university chairs were created
(see Porter, 1981, and Kufbauer, 1982). In France, on the other hand,
chemistry's image was more mundane: like the study of electricity, it was
more a science of amateurs—of experimental demonstrations in salons
or public and private courses—than a university discipline. This may
shed light on the different strategies that were pursued by the Swedes
and Germans on the one hand and the French on the other, to define the
value of their science.

In countries where chemistry experienced a legitimate university development, chemists combatted its image as a crude manual activity. In
this way the distinction between "pure" and "applied" chemistry was
introduced in 1751, by J. G. Wallerius of Uppsala (Meinel, 1983, p. 126).
This distinction fostered a very interesting development: on the one
hand, it asserted the dignity of "pure chemistry" and transformed the
chronological priority of the "chemical arts" into a logical dependence
on the science of which they became simple "applications"; on the other
hand it allowed this "pure" science to benefit from the public utility of
its applications. In France the image of chemistry was linked as closely
as possible to empirical artisanal practices, against the arrogance of those
who wanted to reduce the facts to a "system."

Subversive Chemistry

Macquer addressed his *Dictionnaire de chymie* of 1766 to an "empiricist"
audience. The reader should feel free "to make whatever plan he judges

suitable, and it is very possible that he may make a better choice than the author himself in this regard." As for Venel, like Vicq d'Azyr, he praised only the true chemist who was ready to take off his academic gown and give up the claim of being able to judge without knowing through practice: "It is the necessity of all this practical knowledge, the length of chemical experiments, the assiduous labor and the observation that they demand, the expenses they occasion, the dangers to which they expose us, the eagerness to work at this kind of occupation that one is always in danger of contracting, that has caused the most sensible chemists to say that chemistry is a madman's passion." Becher called the chemist a "certain eccentric, odd, anomalous variety of human who has a very unusual taste that prompts him to lose health, wealth, home and life. But in considering the absolute utility of science as a fact from which the general opinion allows us to reason, these difficulties and drawbacks themselves invite us to consider the 'savants' courageous enough to defy them as citizens who deserve all our thanks" (Venel, 1753, p. 421).

Far from defining the chemical arts as a simple application of a scientific doctrine, Venel defined chemistry as the coexistence of a "double language, the popular and the scientific." "Thus the workman says: a certain degree of heat melts gold, vaporizes water, calcinates lead, fixes nitre, analyzes tartar, soap, an extract, an animal. And science says: a certain degree of heat releases the aggregation of gold, destroys that of water, attacks the mixt of lead and the composition of nitre, excites the reagents in tartar, soap, an extract, an animal" (Venel, 1753, p. 419).[20] The "scientific" chemist allies himself with the artisan against the arrogance of those who deny the mixt because their mechanical agents allow them to account only for the aggregation. He recognizes that a chemical art can never be "reformed" or "improved" except by a chemist who "has acquired that ability to judge by feeling—what is called workman's intuition—which he owes to his close familiarity with his subject."

For Venel, chemistry was unique because it eliminated the social and intellectual division between the scientific chemist and the workman; it fostered the ideal that Diderot advanced (in sections 27–30) in his *De l'interprétation de la nature*. French chemistry, the science of the Enlightenment, was not, however, a "Baconian" science in the purely empirical and utilitarian sense that Diderot has too often been accused of advocating. It was "philosophical," in the sense that it required what Diderot called a "philosophers league" between those who bustle about and those who reflect, between those who "have seen nature at work so often and from so close by" that they have acquired from it "that spirit of divination by which one *guesses* so to speak, unknown procedures, new experiments, unexpected facts" and those who attempt to link up the facts. "Persisting stubbornly in their attempt to solve problems that may be impossible, [they] have arrived at discoveries more important than those solutions."

Diderot announced the end of rational mechanics: "Their (Bernoulli, d'Alembert, Euler, Maupertuis, etc.) works will remain in the centuries to come like the pyramids of Egypt, whose massive façades covered with hieroglyphics inspire in us a terrifying idea of the power and resources of the men who built them" (*De l'interprétation de la nature*, sec. 4). Venel wanted the chemist to keep "his thermometer in his fingertips and his watch in his head" and found ridiculous all the "artificial measurements" people wanted to substitute for the expert reading of "rough and palpable indices." Diderot counseled "those whose spirit is wide enough to imagine systems, *Laidem habeto, dunmodo te Lais non habet,* enjoy the courtesan Lais or the system, but do not let yourself be possessed by her." It is difficult to imagine admonitions more completely refuted by history. Quantum mechanics, our science of atoms, came from the lineage of Bernoulli, Euler, etc. Half a century after Venel, the chemist's laboratory would fill up with instruments for making precise measurements. As for being "possessed" by one's system, that is exactly the definition Thomas Kuhn (1962) gave of the "normal scientist" guided by the paradigm of his discipline.

The history of modern science has not, therefore, fulfilled the expectations of Diderot and his scientific allies. As early as 1780 an extremely astute observer could have predicted the failure of a chemistry of Enlightenment such as that described by Venel. He would have spoken of the new dynamics of academic chemistry and mentioned the names of Guyton de Morveau (member of the provincial Academy of Dijon), Lavoisier, Berthollet, and Fourcroy. While the academic work on the chemistry of salts, some fifty years earlier, had gained for its authors no other reputation than that of empirical chemists, cautious and without ambition, the French academicians at the end of the eighteenth century were a new embodiment of progress. They were part of a European network whose work gave unprecedented meaning to what in Germany and Sweden was already called "pure chemistry."

But no doubt this same astute observer would have been mistaken as to the way in which the pure chemistry of the academicians would finally succeed in wiping out the memory of the great themes of Enlightenment chemistry. Presumably he would have spoken of the very exciting correspondence between two great chemists, Guyton de Morveau of France and Torbern Bergman of Sweden, working together on the interpretation of affinities in Newtonian terms. He would have described the triumph of what Voltaire had nicknamed the "Newtonian party in France," led by its "chief," the great naturalist Buffon. But he probably would have been deaf to the "revolutionary" message of what had begun less than thirty years earlier as a soft sound, the fizzing of the lime on which a Scottish hand had poured a few drops of acid. Before listening to the unexpected revolution, let us describe the failure of the anticipated one.

The Newtonian Dream

Can chemical relationships or affinities enable us to identify the "forces" responsible for chemical combination? That is a question one would expect to have interested the English chemists. But it did not. Boyle's skeptical pragmatism flourished in the soil of England, where the Royal Society encouraged measuring the value of a science by its utility. It was rather on the continent that the "Newtonian dream" of subjecting chemistry to Newtonian forces in an effective way would develop. (See Thackray, 1970.)

Understanding Affinities

English skepticism and the continental dream were fed by two analogous theses about the possibilities of reconciling specific chemical affinities and the uniform gravitational force. The theologian Ruggiero Boscovich (1785) assumed that Newton's formula applied only to large distances; at short distances the force could be either attractive or repulsive, depending on the distance: a chemical body, a complicated structure of particles, was therefore specified by the force resulting from these attractions and repulsions. Buffon (1765) attributed the specificity of chemical reactions to the shapes of the bodies, which, on the scale of chemical activity, could no longer be neglected, as in astronomy. The consequences of these analogous theses were different, however. Boscovich thought that the complexity of the structure of any chemical body made chimerical the prospect of a deductive and predictive science like astronomy. Buffon, on the other hand, thought his "great grandchildren" would one day calculate the progress of chemical reactions as Newton had calculated that of the planets.

In Buffon's perspective, chemistry was not yet a "true science." It would become one when it actually operated on the model of Newtonian astronomy. This brings up the question of affinities. How to transform the tables of affinity—which could, in this respect, be compared to the empirical facts of observational astronomy, before Kepler and Newton formulated their laws—into a truly scientific chemistry? During the last three decades of the eighteenth century, research inspired by the desire to make affinities the object of a science was initiated, and these austere works take us back to the studious atmosphere of academic and university exchanges.

In 1776 Guyton de Morveau, enthusiastically following the program outlined by Buffon, measured the mechanical force necessary to separate plates of different metals from the mercury bath in which they were floating. Guyton was trying to quantify affinity, to measure the relationship between two bodies independent of the chemical operations of displacement. Indeed, displacements could produce only a relative measurement (i.e., characterize bodies in relationship to each other).[21] Along the same lines, Carl Wenzel, a chemist working in a porcelain factory, proposed another method in 1777, which linked the numerical value of the affinity to the speed of reaction. He determined the weight loss of metallic cylinders which had been placed in an acid bath for one hour. Then Wenzel, Bergman, and Richard Kirwan used as a measurement the respective amounts of different acids needed to neutralize the same amount of base, or, inversely, the amounts of different bases needed to neutralize the same amount of acid.[22] A similar work was undertaken on a larger scale by Jeremias Richter, a former student of Kant intrigued by the possibilities of a quantified chemistry. Between 1792 and 1802 he published, in a form that his desire to mathematize makes rather obscure, a new type of table systematically defining the relationships, which he called "stoichiometric," between mutually neutralizing acids and bases. The term *stoichiometry* would be applied not to the field of chemical affinities but to analytic chemistry, which is based on the law of definite proportions.

An apparently more modest path was to use the method of the "table makers," as they were called, not as a condensed and convenient way to present chemical knowledge but as an end in itself. Since the affinities were an empirical subject meant to be organized into a deductive scheme someday, this area had to be explored methodically and *exhaustively.*

Bergman's Tables

Studying all possible chemical reactions in order to put them into tables was the project undertaken by the Swedish chemist Torbern Bergman. Published between 1775 and 1783, his tables imply an enormous amount of tedious work. They put several thousand chemical reactions in order. They contained 49 columns (27 acids, 8 bases, 14 metals, and others) and discriminated reactions by the wet method (in solution) and the dry method (by fire). As Maurice Daumas pointed out, Bergman worked "as a conscientious artisan who hoped to nibble away little by little at the immense work that extended before him . . . He apparently did not consider his own results very satisfying: he estimated that more than thirty thousand meticulous experiments were still necessary to give a certain degree of perfection to his table" (Daumas, 1946, p. 61).

Bergman's gigantic labor included work on chemical nomenclature, using Linnaeus's methods, which he had discussed at length with Guyton de Morveau (Beretta, 1988). It was also accompanied by a symbolization of chemical reactions that suggested they be understood in terms of the associations and dissociations of those constituent parts that remained untouched. The components were represented, as had already been done in Geoffroy's table, by symbols deriving from alchemy. The compound was placed outside a bracket that united the components. In the case of a double decomposition, four brackets were arranged into a square, with the two horizontal ones corresponding to the two products of the reaction. Bergman also indicated the solubility (the point of the bracket up) or insolubility (point down) of the substances.

The study of all possible chemical reactions illustrated the consequences of the relational conception of affinity inspired by Newton. Chemical bodies as such, or the power of this or that reagent, were no longer interesting. The power of nitric acid, for instance, could no longer be illustrated by some typical reactions and then defined as an attribute of this acid. Chemical properties depended on relationships among bodies, and the identity of a chemical body would be defined by the complete set of its possible relationships, as long as deductions remained impossible. With the chemistry of salts in the first decades of the eighteenth century, reactions, which in the artisanal tradition were used to create new products, became an instrument for identification, but they still concerned products—acids, salts, bases—that were interesting for practical as well as cognitive purposes. With Bergman the reaction is no longer the means; it had become the *phenomenon*. Henceforth every reaction would be interesting.

Because he had to be interested in every reaction and not only the "interesting" ones, Bergman encountered problems that other, less ambitious table makers had ignored. (For what follows, see Holmes, 1962.) What exactly is an interesting reaction? It is a reaction that produces a new compound. There was nothing fortuitous about the stress placed on the distinction between "fixed" and volatile bodies in alchemy and then in chemistry: most of the reactions then used produced either one or the other, for, in both cases, the product left the reactive environment (which, in modern terminology, brought on a complete reaction). Moreover, the prevalence of complete reactions enables us to understand one of the essential aspects of the "mixtive union": if one body had a superior affinity for another body than that which united the second to a third, the first *completely drove off the third* (if it existed in sufficient quantity, of course). All chemists, whether Newtonian or not, admitted this, and Bergman referred to it again in speaking of "elective attractions." The elective characteristic was essentially different from the simple mixture or aggregative union. The elective affinity reflected a tendency to unite according to an all-or-nothing logic: the strongest wins.

Examining chemical reactions as a whole, Bergman encountered many anomalies and had to multiply the distinctions between the "truly chemical" affinity and the physical factors that were obstacles to it (i.e., factors that prevented the displacement from being completed). He even realized that sometimes a reaction goes in one direction, sometimes in the other, according to the quantities of reagents available!

If there is a good case of "paradigm shift" (Kuhn's expression) here, it is certainly the way in which Berthollet, former collaborator of Lavoisier, would transform the anomalies that proliferated with Bergman and others into laws. But the paradigm shift did not mean that the

Newtonian dream had been abandoned, rather that it had finally been put to work. A half-century later Berthollet would finally prove Venel correct: the Newtonian force of attraction could not account for the difference between chemical combination and mixture. As a consequence Berthollet would deny this difference.

A Newtonian Chemistry at Last?

Claude-Louis Berthollet participated in history during an extraordinary period. Academician, industrial chemist, traveler, teacher, senator, founder of the Society of Arcueil, the first "interdisciplinary group," he had an agenda that if it does not explain at least clarifies his uniqueness.

During the French Revolution, Berthollet sought to standardize the extraction of saltpeter (needed to produce cannon powder). Giving up the old method of washing the nitrous rocks on site for the sake of centralized production allowed him to make a strange discovery: it was better to wash the rocks several times, using fresh water each time, because the more saltpeter was already dissolved in the water, the less effective was the washing.

Perhaps many artisans had made this observation, but Berthollet translated it, as Venel would have said, into scientific language: the tendency of a body to combine with another decreases in proportion to the degree of combination already reached. The tendency to react, from that time on, was no longer seen as purely chemical, but as a *function* of the concentration of the reagents. As early as 1795 Berthollet asserted this in his course at the new Ecole Normale. Unlike the academicians of the ancien régime, the chemist now had the duty to deliver a "course." The new scientific lecture was no longer addressed to the curious, enlightened public, apothecaries, or artisans. It was meant to educate future teachers. It could be read critically by other colleagues, so it was necessary for the author to convince people like Laplace, Lagrange, and Monge, who knew what it meant to reason. Perhaps without this new challenge Berthollet would not have had the courage to break with tradition.

In 1800 Berthollet accompanied Bonaparte to Egypt, and it was there, it is said, that he made the observation that would precipitate his "conversion": he encountered a "lake of sodium"—that is, a salt lake whose banks were covered with "soda" (to which, according to the reformed nomenclature, we can give the modern name of sodium carbonate).

Now Berthollet knew as well as everyone else that in the laboratory salt does not transform spontaneously into soda: the spontaneous reaction goes in the opposite direction. But this lake was a very special reactional environment: the two reagents, the salt and calcium carbonate deposited on the bottom of the lake, were abundant, and the two products of reaction were continuously eliminated—calcium chloride drained into the soil, and sodium was deposited on the banks.

Berthollet drew a radical conclusion from this example: the direction of a reaction was not an *absolute*, determined by the elective tendencies of the bodies present. The distinction between a truly chemical phenomenon, which corresponded to a complete reaction, and the physical factors that explained anomalies should be abandoned. Berthollet thus reversed the terms of Bergman's judgment. For him, it was complete reactions that must be explained. Using Bergman's tables, he showed that each time the displacement was complete, a product escaped from the reactional environment, either because it was volatile or because, not being very soluble, it precipitated out.

Back in Paris Berthollet became one of the "great men of science" honored by Napoleon. Like Laplace, he would be a senator and, with Laplace, he would found the Society of Arcueil, a private society with its own laboratories, journal, and, above all, discussions and collaborations (Crosland, 1971; Sadoun-Goupil, 1977). The experimenter was no longer alone, and the milieu in which he worked had a doctrine: physical and chemical processes could be explained by Newtonian forces of attraction. In his *Essai de statique chimique* (1803), Berthollet showed that the Newtonian concept required that chemical reactions be generally incomplete and that physical conditions (temperature, concentration) not be defined as "permitting" or "preventing" the properly chemical action of affinities. Affinity was in fact only one factor among others in the reaction, which no longer had a natural direction but one determined by a whole set of conditions. The normal result of a reaction was an equilibrium in which the relative concentration of the reagents and products of reaction was a function of this group of conditions (from which came the term *statique chimique* in the title). The only peculiarity of affinity was that of the force of attraction: it could not be manipulated directly by the chemist.

So Berthollet had found a third way to make chemistry a Newtonian science: the example to follow was no longer astronomy but mechanics, which had also begun with statics, the study of equilibrium. But as the

Newtonian dream was finally considered with all its consequences, it turned out to be a nightmare for the chemist. Venel was right: the force of attraction could not explain a complete reaction; worse, it could only account for a mixture, not a real combination (i.e., the actual production of a new homogeneous body from a heterogeneous one). Every compound was therefore a mixture whose composition depended on the reaction conditions.

Here Berthollet had struck a blow to the practical identity of chemistry. If he was right, all the procedures that were the triumphs of this science not only rested on specific cases but simply did not accomplish what they claimed to do. The chemist who had learned to identify a body by its composition was mistaken. A chemical compound did not have a fixed identity; each compound was a particular mixture that depended on the conditions of production. Elective affinity, interpreted in strictly Newtonian terms, led chemistry to a generalized skepticism, as corpuscular chemistry had done in Boyle's time. Once again the only valid theory for chemistry had to come from physics.

Controversy over Definite Proportions

Usually when the controversy raised by this theory, which would last until 1807, is mentioned, Berthollet's adversary, Joseph Louis Proust, is seen as an advocate for the future. Proust is portrayed as a modern chemist who established experimentally, in spite of the authority of dogmatic speculation inherited from the eighteenth century, what we know to be true: chemical compounds certainly do have an identity, with constituents in fixed and *definite proportions*. In fact the controversy can be read differently: it was the "revolutionary" proposition that was vanquished, while traditional chemistry carried the day.

Before engaging in an experimental controversy with Berthollet, Proust had published a first article in 1799 in which he showed that oxide of mercury always had the same composition, whether it was produced in the laboratory or mined in Peru. At the time Proust did not consider this a revolutionary conclusion at all. It was a confirmation of Stahl's old doctrine of the invariability of *pondus naturae*: the definite proportions by weight of compounds suggested that the invisible hand of nature operated identically in the depths of the earth and in laboratories. As for the opposition between fact and speculation, both protagonists appealed to facts that were equally ambiguous. Both analyzed metallic oxides, and we know in retrospect that the experimental techniques of the day could not separate them: what they analyzed was, in fact (i.e., for us), a mixture of oxides of different types. It was also Proust's interpretation, but Berthollet had no reason to accept it and accused his adversary, with good reason, of using ad hoc explanations to mask the fact that any body was a mixture, only a mixture.[23]

The controversy petered out into general indifference. Neither protagonist saw himself as the loser, but in the meantime the law of definite proportions had become a "fact" for all chemists. In 1832 Gay-Lussac would draw "our" moral from the story: Berthollet was right about reactions—they generally produced an equilibrium among the constituents, which depended on the reaction conditions. He was wrong about the compounds produced. The Newtonian dream was dead—namely, the ambition to determine relationships that would explain the bonds and the process at the same time was dead. The two problems were now separated. The chemist could manipulate the reaction, which was a function of the temperature, the concentrations, and other physical factors. He could not manipulate the bonds themselves, which assured the individuality of chemical bodies. Soon Davy, Faraday, and Berzelius would suggest explaining the bonds by electric forces. That attempt would fail, but the stakes had changed in any case: neither this proposition nor its abandonment was enough to put the identity of chemistry into question. Chemistry had become a science, but not through the accomplishment of the Newtonian dream.

What is more, the notion of chemical equilibrium would not interest anyone for decades to come. As Wilhelm Ostwald pointed out, the chemistry of the first half of the nineteenth century turned back to the traditional interests that made the chemical reaction an instrument, not a phenomenon in itself: "Even in new areas, people applied themselves especially to the preparation of bodies. The conditions of a reaction were always fairly well studied when they had found a good procedure for obtaining this or that body, and nobody was interested in studying in depth the bad methods that did not result in pure products" (Ostwald, 1906, p. 211).[24] Again they were interested only in complete reactions.

The death of the Newtonian dream can be read as the confirmation of Venel's claims and of all those who, with Stahl, defended chemistry against the ascendancy of mechanical agents. But it was not the French chemistry of the Enlightenment that triumphed, for the controversy in which the Newtoniam dream died took place in the framework of post-revolutionary chemistry. Just as Berthollet did, those who drew the lessons from it defined themselves as teachers and servants of the state. In 1807 Jean Antoine Chaptal published *Chimie appliquée aux arts*, in which he deduced the truths of "pure," autonomous chemistry from "the art that should be their useful result." Academic space would hence-forth be represented as closed: no one saw the future as a dynamic alliance between artisans and "savants" anymore, and Venel's criticism of the usefulness of measuring instruments in the practice of chemistry was no longer audible at all. What is more, the chemistry of principles was dead, and references to Stahl's doctrines were seen as irredeemably out of date. Of course, Lavoisier had stricken. But ironically, his inter-vention was itself conditioned by the unexpected fecundity of one of Rouelle's four principles, the fire-phlogiston that would accompany the accelerated development of pneumatic chemistry.

Hunting "Airs"

Pneumatic chemistry was born on the margins of the strongholds of Newtonian doctrine, in the laboratory in Glasgow where Joseph Black (1728–1799) prepared a doctoral dissertation under the supervision of William Cullen (1710–1790). (For what follows, see Donovan, 1975.) William Cullen, who had studied in London until 1740, had seen how Newtonian chemistry developed after the death of Newton (1727). The first priority for English chemists was the confirmation of the hypotheses advanced in "Query 31." Thus, in *Vegetable Staticks*, Stephen Hales pre-sented his experiments on the fermentation of vegetables as a demon-stration of the Newtonian proposition that the "true permanent air" given off by fermentation, or by heating bodies that chemists call "fixed," consisted of particles that were mutually repulsive. This "air" obeyed Boyle's law of compression; put another way, it confirmed the generally elastic nature of air. For Hales, elastic air was in itself heterogeneous, consisting of particles of different elasticities. The least elastic particles could be fixed the most easily and were the ones obtained by heat or fermentation. Hales thus differentiated among the airs in physical terms. For English chemists and physicists, air was first of all the means of characterizing the repulsive force, about which the laws of motion were mute. Until the 1770s, they would study the new electrical and magnetic phenomena from the same perspective.

Newton's principal heir among the chemists, Desaguliers, denied the existence of substantial fire and defined heat in terms of the repulsive force between particles. In 1748 William Cullen presented Desaguliers's theory as an example of the kind of error that demonstrates that chemists must formulate their own theories rather than apply the general mechanical laws to their field. Cullen believed fire was attracted in a *specific* way by different chemical bodies. He showed by quantitative measurements that expansion, contraction, and the temperatures necessary for state changes (which were seen as effects testifying to the presence of fire in a body) followed a *specific* law. So Scottish chemists returned to the problem of specificity, unlike the Newtonians in London. Cullen was already speaking of "elective attractions" and circulating copies of Geoffroy's table among his students.

A Scottish Chemistry

The freedom with which Cullen treated the Newtonian doctrine no doubt owed much to the dissident voices that had dared to make themselves heard in London after Newton's death. Perhaps Cullen took the course of Peter Shaw, who in 1730 translated Stahl into English and thereafter realized gradually that defending the autonomy of chemistry meant questioning Newtonian chemistry. But Stahl's chemistry would not satisfy Cullen's need for demonstrations. The University of Glasgow's laboratory was no longer the typical laboratory of the seventeenth century. There were thermometers, air pumps, and pneumatic chests for collecting "airs" and other products of evaporation. It was a good place to wonder and to discover why, for example, when alcohol evaporated, the temperature decreased, and decreased still more if evaporation took place in vacuo. Or to ask why a mixture of lukewarm water and ice caused the ice to melt without raising its temperature. In fact the Glasgow laboratory was in a rather unique situation—at the intersection of Paris, with its salts and its phlogiston, and London, with its forces and its physical chemistry of "airs."

It was known that when acid was poured onto calcareous earth, it effervesced, and that calcined limestone produced a caustic quicklime

(CaO)—this was even its defining property (*calcareous* derives from *calcination*). But as we saw with the French chemistry of salts, definitions by properties were no longer reliable: effervescence no longer guaranteed that there was but a single limestone. Perhaps each calcareous earth produced a different quicklime when calcined? Had not Duhamel shown that the "base" of marine salt and that which could be derived from tartar salt are different? And how to understand *magnesia alba* ($MgCO_3$), which had the effervescent properties of calcareous earth but which produced a substance that was neither soluble in water nor caustic when calcined? This was the question that Cullen proposed to his pupil Joseph Black in 1752 as the subject of his doctoral dissertation in medicine. Black did not like medicine, but his father wanted him to be a doctor. He was fond of chemistry, and the choice of *magnesia alba,* that useful purgative, was a happy solution, all the more so since its differences from and similarities to the calcareous earths justified his studying them as well.

For three years Joseph Black undertook a series of experiments whose results were "positive" in the sense that he would later be able to use some of them to state the problem, others to demonstrate his thesis, and others as confirmations of consequences predicted by it.

Fixed Air, Fixed Heat

Calcination leads to the release of fixed "air" in calcareous earths as well as in *magnesia alba.* Limewater (caustic) in the presence of caustic potash could "take back" the fixed air—regain its original state as calcareous earth—which meant losing its caustic character as well. To regain the original *magnesia alba* from its calcined product (noncaustic), one had first to dissolve it in spirit of vitriol; the addition of tartar salt would then cause the *magnesia alba* to regain all its original properties. The same amount of acid was necessary to "saturate" (neutralize) a given amount of limestone and the quicklime produced from that amount of limestone.[25] As a result of these experiments—Black mentioned thirty in his dissertation, each presented as a confirmation or refutation of one possible interpretation—the author could announce that the causticity of quicklime was not, contrary to what had been believed, determined by a substance extracted from the calcining fire; rather, it was an intrinsic property of calcinable substances (which, for an unknown reason, *magnesia alba* did not possess). Causticity was defined as an elective attraction for certain substances that quicklime dissolved or corroded, an attraction that was imperceptible when the air was fixed in the stone.

Black knew that the air fixed by lime was a specific kind of air; he knew that it was unbreathable and did not enter into combustion, fermentation, or respiration. But that did not interest him. What did interest him was to be able to submit fixed air to displacement, and to be able to add a column to Geoffroy's table with fixed air at its head. After the publication of his dissertation in 1756, Black abandoned calcareous earths to study the question of heat, which also fascinated Cullen. This was an extension, not a change of subject, for Black would produce a real "chemical theory of heat."

Not knowing what "fire" was, Black avoided speaking of it, but he studied the specific relationships between two measurable quantities, heat and temperature. The distinction between quantity of heat and temperature was not new in itself. It was first used by Guillaume Amontons when he was making an air thermometer; it was, in fact, brought about by the very question of the thermometer: it was necessary to define the quantity of heat separately, since it was as a function of this quantity that the thermometric body had to dilate in a linear way. Black was interested in the way in which heat was "fixed" in bodies. He called this "latent heat." Latent heat was the heat absorbed or released by a body during a change of state without a change in temperature. Contrary to the usual absorption of heat, the absorption of latent heat is not comparable to that of a fluid in a porous body. It must be understood as a *combination* that entails melting or boiling.

According to legend, Black's explanation of latent heat to the young James Watt was the source of the invention of the steam engine (see Cardwell, 1971). It is clearer that latent heat, heat "fixed" within bodies, provoking a change in the state of aggregation of these bodies and losing its property of heating, was not unrelated to "fixed air," which also lost its property of volatility and caused a change of state, the disappearance of the causticity of quicklime. Since causticity and cohesion were both related to attractions, one may, with Donovan (1975, p. 215), conclude that Black had less founded pneumatic chemistry than pursued the first research in theoretical chemistry, searching for "general laws that link heat, elective attraction and changes of state." Heat and its measurement, which Venel had rejected as belonging to physics, were integrated by Cullen and Black into the economy of chemical transformations.

Nevertheless, Black was a pioneer of pneumatic chemistry. He submitted an "element-instrument," air, to the chemistry of displacement. After air, it was phlogiston that would be displaced from body to body. In 1766, Henry Cavendish (1731–1810) isolated and defined the "inflammable air" (our hydrogen) given off by metals attacked by acid. Such a release had been reported by Helmont and Boyle, but unlike them, Cavendish was able to identify it. Could not what he had isolated be the famous phlogiston in its pure state, which would therefore be shown to be a principle both of combustibility and of metallicity?

European Competition

So the race for different airs was on, under the aegis of phlogiston. It spread across all of Europe, from Uppsala to Paris, London to Berlin, at a time when European chemists were creating a real communications network through an exchange of correspondence, visits, and journals. Lorentz Crell's *Chemische Annalen* solicited contributions not only from German chemists but from all over Europe. It united chemists of different status: Cavendish, like Boyle, was a wealthy member of the nobility; Lavoisier was an academician and administrator who equipped his laboratory at his own expense; Joseph Priestley (1733–1804) was a Unitarian minister to whom a noble protector had donated a laboratory; Karl-Wilhelm Scheele (1742–1786) was a self-taught apothecary who worked in poverty and obscurity without institutional affiliation and who wandered from Stockholm to Uppsala and from Uppsala to Köping, where he finally bought an apothecary shop. Their preoccupations were also very different. Cavendish was timid and eccentric, methodical and reserved: "His head seems to have been but a calculating machine; his eyes, inlets of vision and not fountains of tears," wrote his biographer (cited in French, 1941, p. 155). Scheele was above all a brilliant discoverer—he prepared, identified, and studied dozens of new substances. Priestley's interests went from chemistry and electricity to the study of languages, theology, history, politics, and the improvement of technology for human welfare. He shared this last preoccupation with Lavoisier, who, in addition, was contemplating a new foundation for chemistry.

In Scheele's laboratory more than fifteen new acids were identified (notably hydrocyanic acid, whose taste and smell Scheele described). For the first time the acrid fumes of chlorine, which Scheele identified as the acid of dephlogisticated sea salt, were given off. Phlogiston also allowed him to define one of the two "airs" that formed the atmosphere as *Feuerluft*, "fire air." *Feuerluft* was capable of absorbing the phlogiston from bodies, like iron filings, that were rich in it, and its union with phlogiston allowed it to escape from the container, which then held only *Verdorbeneluft*, the corrupt, vitiated air that mice died in. *Feuerluft* also disappeared when Cavendish's phlogiston burned in air, thus obviously generating fire-heat. On the other hand, this same *Feuerluft* was produced by reducing through calcination a phlogiston-thirsty body, such as calx of mercury. The fire released its phlogiston and the residue was nothing other than *Feuerluft*. The facts were consistent (except the weights, which did not interest Scheele): *Feuerluft* was an air that produced fire-heat when united with phlogiston.

In parallel with Scheele, an English chemist also identified "our" oxygen in 1774. Joseph Priestley, author of several volumes entitled *Observations and Experiments on Different Kinds of Air*, systematically collected the gases released by reactions in a mercury trough. He had already isolated and identified in the gaseous state what we call hydrochloric acid, ammonia and sulfur dioxide, and he also discovered hydrogen sulfide, hydrogen phosphate, ethylene, and our nitrogen, which he called "phlogisticated air." The uncontested champion in the search for airs, Priestley was also a pioneer in studying living things chemically. Each air that he isolated was characterized by a certain number of reactions. One of his tests for identification was to put a plant or animal under a bell jar filled with the new air to be defined. He advanced the chemical study of respiration along with the study of gases (Holmes, 1985).

Even if respiration was only one way among others to serve an end—namely, to define new airs—that did not prevent Priestley from making hypotheses on the nature of the mechanisms at work. So it was that from 1771 on he likened animal respiration to a kind of putrefaction: a phlogistification that corrupted the air. He suggested that the role of the lungs was to evacuate putrid emanations toward the outside, and he showed that plants were capable of "restoring" the air and making it breathable again. The idea of corruption caught Priestley's interest and that of some of his colleagues in England and Italy, who were laying the foundations for a scientific study of the salubrity of the air in our environment. When Priestley isolated oxygen, however, by reducing an oxide or calx of mercury, then called *mercurius precipitatus per se,* he did not immediately define it with the respiration test: he declared that it was soluble in water, that it made the flame of a candle placed in its container burn brighter, and he at first took it for an unbreathable nitrous gas.

It was on the nature of this new air that Lavoisier intervened, and thanks to oxygen he would be recognized as chemistry's great reformer. During a trip to Paris in October 1774, Priestley spoke to Lavoisier about his experiment on the reduction of mercury calx. Lavoisier, who was also acquainted with Scheele's work, redid Priestley's experiment as a way of solving the following problem: "Are there different types of air? Is it enough that a body be in a state of expansion to form a type of air? Are the different airs that nature offers us or that we manage to form discrete substances or modifications of the air in the atmosphere?"[26] In fact, Lavoisier did not really settle the question in those terms in 1775, any more than Priestley had. He was not then able to recognize the oxygen he had prepared. He did reveal some doubts on the name "nitrous air," which Priestley had given it, but he was still describing it as "common air," the stuff we breathe. It was only after Priestley, going back to his first interpretation, renamed it "dephlogisticated air" that Lavoisier re-did the experiments on the reduction of mercury calx and firmly concluded that "dephlogisticated air is the purest part of atmospheric air." He then began a polemic against Priestley, attacking his theories of combustion and respiration in two papers published in 1777.

Who, then, discovered oxygen? Scheele, who first isolated it? Priestley, who defined its properties? Lavoisier, who identified it as an element?

Water presents the same problem of priority. In 1781 the very meticulous Cavendish, burning his "inflammable air" in Priestley's dephlogisticated air, noticed the formation of a damp mist on the glass bell jar. So Cavendish proposed to redefine Priestley's air as dephlogisticated water. Priestly objected: it was Cavendish's "inflammable air" that was water overcharged with phlogiston, which during combustion was given up to dephlogisticated air.

Lavoisier had been trying for several years to demonstrate experimentally that burning inflammable air (hydrogen) with pure air (oxygen) yielded an acid, and since he had not succeeded in doing it, he decided to have a large combustion apparatus with two gasometers constructed. Then he learned of Cavendish's results from his assistant Charles Blagden, who visited Paris in 1783. He redid the experiment and drew the conclusion that water consisted of inflammable air and pure air. In brief, Lavoisier changed the interpretation of the phenomenon discovered by his colleagues. Using their procedures, he proposed a different theoretical reading of all the experiments. Until then his colleagues could take Lavoisier for a "hunter of airs," a member of the pack, like themselves. But in the 1780s he used the development of pneumatic chemistry to serve another objective, a new project. All the specimens of gas collected and identified before him would have to be renamed and redefined without reference to phlogiston, as elements making up atmospheric air and water.

How was this theoretical alternative conceived? Why eliminate phlogiston at the very moment it had acquired an experimental reality, thanks to Cavendish's identification of it with his inflammable air? How would chemists, who had been struggling tooth and nail in an international hunt for various airs, react to the news? What did they make of this upset, which was called, on the suggestion of Lavoisier himself, a "revolution" in chemistry?

A Revolution in the Balance

Lavoisier began to wonder about the nature of phlogiston in 1772, while exploring calcination. (For this section generally, see Guerlac, 1961, 1975.) He was working on a problem that had been known for a long time but had recently been discussed by Guyton de Morveau: if, as it was then thought, the calcination of metals (what we call "oxidation") involved a release of the phlogiston contained in the metals, why was there a weight gain in these metals? Guyton suggested that since phlogiston was lighter than air, its presence in a substance made that substance lighter.

Crucial Experiments?

After two experiments Lavoisier attempted a different explanation. He burned sulfur and then phosphorus in closed vessels and noticed, by scrupulous weighings of each and every part before and after the reaction, that the total weight was conserved: the weight of the vessel was unchanged, while that of the sulfur and phosphorus had increased. He concluded: "This weight gain comes from a prodigious quantity of air that is fixed during the combustion and combines with the vapors." Convinced of the revolutionary importance of this experiment and anxious to control its interpretation, Lavoisier sent a sealed note to the Academy on the first of November, 1772, to assure himself priority in a discovery that he judged "one of the most interesting that had been made since Stahl." It is a great temptation to see this theatrically presented experiment as the founding act of modern chemistry. Using the balance and the famous principle that "nothing is lost, nothing is gained," Lavoisier turned the phantasmagorical phlogiston on its head and with it the whole doctrine of the four elements. Lavoisier himself would favor this reading of his discovery later on, during the controversy over phlogiston.

If, however, the situation is described not only through Lavoisier's proposals but also through those of his colleagues and contemporaries, we become inclined to reevaluate the impact of these first attacks against phlogiston. First of all, the weight gain of calcined metals was not an anomaly or a defect discovered by Lavoisier. It had been, as the academicians' report on the works of Lavoisier recalled, well known since the seventeenth century and had received various explanations, at least two of which were very close to that of Lavoisier. Two essays by Jean Rey and John Mayow would be recognized as precursors of Lavoisier, but they had remained unknown as long as the weight problem had not been considered crucial.

Second, Lavoisier was neither the first nor the only one to criticize phlogiston. In 1773 and 1774 two anonymous articles appeared in the *Journal de l'abbé Rozier* denouncing the fictional and hypothetical character of phlogiston in almost the same terms that Lavoisier would use ten years later. And a first campaign against phlogiston was launched by Buffon in 1774.[27]

Third, Lavoisier did not yet have the means—nor even the intention—of overthrowing phlogiston in 1772. If we consider the whole of his work—about fifty articles read at the Academy, the *Méthode de nomenclature*, the *Traité élémentaire de chimie* (without mentioning here his economic and financial works)—the question of phlogiston does not appear primary.[28] In our reading of Lavoisier's works, we should therefore exercise the same caution with which one should approach pre-Lavoisier history—not to read it backward in order to avoid distortions and reductions.

We should simply emphasize that Lavoisier was working on the analysis of gypsum and of mineral waters, combustion and calcination, the nature of acids, the fabrication of saltpeter, the respiration and transpiration of animals, aeriform fluids (gases), the theory and measurement of heat, the composition of water and air, the affinities of the "oxygen principle" ... a great variety of subjects dominated by his interest in gases.

Also, in joining the main current in chemistry in the 1770s, Lavoisier provoked many controversies because he distinguished himself from his English colleagues by a tendency toward theory. He advanced theories on combustion, the nature of acids, the gaseous state, respiration, and transpiration. He was always inventing hypotheses in a continual seesawing between theory and experiment, between his own experiments and those of others. Even though, in polemics, everyone appealed to the power of facts, even though Lavoisier later proclaimed the death of phlogiston in a solemn condemnation of hypotheses and systems, it is clear that his professions of epistemological faith were polemical formulas directed against a very precise target and of circumstantial application.[29] Moreover, the victory of oxygen over "dephlogisticated air" was less a modification of Priestley's experimental results than a change in their significance: the object was no longer the pursuit of yet another specimen in the hunt for different airs; now it was a tool for analytical research. By shifting the focus of the international competition in gases in this way, Lavoisier took leadership of the movement.

Then he felt confident enough to launch a first attack against phlogiston in a memoir, *Sur la combustion en général*, presented at a public session of the Academy in 1777. This lecture was awaited as a great event. It was rumored that Lavoisier was going to overturn Stahl's doctrine. But the audience was rather reassured. Macquer expressed his relief: "Mr. Lavoisier has long been frightening me with the prospect of a great discovery, which he was keeping secret, and which would do no less than turn upside down the whole theory of phlogiston or combined fire: his air of confidence made me tremble with fear. Where would we have been with our old chemistry, if it had been necessary to rebuild a completely different edifice? For myself, I assure you I would have abandoned the game. Happily, Mr. Lavoisier has just disclosed his discovery in a memoir read at the last public assembly; and I assure you that since that time there is a great weight off my chest. According to Mr. Lavoisier there is no matter of fire in combustible bodies; it is only one of the constituent parts of air; it is air and not what we regarded as a combustible body that is decomposed in all combustion; his igneous principle is given off and produces the phenomena of combustion, and there is nothing left but what he calls the base of the air, which he admits is entirely unknown to him. Judge for yourself whether I had reason for such a great fear" (Guyton de Morveau, 1786, vol. 1, p. 628).

Was this blindness in the face of Lavoisier's new theories? We must recognize that at that date Lavoisier had not really overturned but merely displaced the phlogiston from the combustible into the air. It was a simple inversion of the combustion scheme: instead of a decomposition yielding phlogiston, there was a combination with one part of the air releasing the matter of fire.

This inversion was not enough to create a distinction between Stahl and Lavoisier. Several chemists proposed compromises. In the second edition of his famous *Dictionnaire,* Macquer accepted Lavoisier's conclusions: the necessity of air to produce combustion, the decrease in weight of the air, and the increase in weight of the combustible. He suggested that air replaced the phlogiston contained in the combustible body. Macquer was much more disturbed by Buffon's attack over phlogiston in 1776, which led him to revise his own theory. He no longer defined phlogiston as matter of fire but as light. In 1782 Richard Kirwan, like Cavendish, identified "inflammable air" (hydrogen) with pure phlogiston, and he proposed another compromise: the phlogiston emitted during combustion combined with the dephlogisticated air (oxygen) to form fixed air, which was present—according to him—in calces and acids, thus explaining the weight increase (Kirwan, 1780, pp. 232–233). Today these theories of double solution, adopted by the majority of Kirwan's contemporaries, seem as vain as the explanation imagined by Tycho Brahe to reconcile Ptolemy and Copernicus. They reveal, nevertheless, that neither the famous experiment of 1772 nor Lavoisier's theory of combustion and calcination elaborated during the five following years was sufficient to overthrow phlogiston.

The attacks on phlogiston would become significant only in a larger context: the theory of gases that Lavoisier developed throughout his work.[30] The key actor in this theory was the caloric, the substance of heat or matter of fire, which crept in among the constituent parts of a substance and gave it expansibility.[31] If the physical state of a body—solid, liquid, or gaseous—were explained by the quantity of caloric that it contained, air lost its essential function as a principle. Caloric also struck a blow in the battle against phlogiston, because it allowed Lavoisier to explain the production of heat or light in combustion: the fixation of oxygen liberated the caloric that was united with it in the gaseous state. Although caloric differed from phlogiston because it could be measured with an apparatus (the "calorimeter") designed by Laplace and Lavoisier, it was nevertheless an imponderable element with its own properties. Lavoisier did not therefore define all chemical elements as ponderable. He adopted Cullen and Black's substantialist conception of heat, but, as Boerhaave had formerly done, he gave it repulsive effects.

The caloric was not the only element-principle in Lavoisier's system. Another essential agent was oxygen—responsible for combustion and calcination and bearer of acid properties, which earned it its name. Oxygen exhibited the double behavior of Rouelle's elements—as universal constituent and agent of reactions.

So then how did Lavoisier convince his contemporaries that he was creating a revolution and convert them to his views? Water seems to have been the decisive element in this affair. Although Cavendish's experiment had been verified by Monge, Lavoisier decided to do it again with his new apparatus, and he transformed the demonstration of water's composition into a historic national event. On the twenty-fourth of June, 1783, the king, a minister, the English chemist Charles Blagden, and some academicians took their places in front of the combustion apparatus to witness and agree that Laplace and Lavoisier, having turned on the faucets for the two gas reservoirs, collected several drops of water in the tube of a funnel.

At that time the balloons developed by the Montgolfier brothers and the physicist J. A. C. Charles were arousing great interest, and the king asked the Academy to perfect the system. Two years later Lavoisier, who had been given the task of producing inflammable air, was able to give an even more spectacular demonstration of the composition of water, as the result of a big experiment requiring analysis and synthesis that lasted two days. Determining the nature of water mobilized more financial, technical, and human means than that of air. But it was followed by the "conversion" of some of Lavoisier's colleagues to his theories as a whole. After twelve years of research that multiplied doubts on the foundations of the chemistry of elements, it was a "drop of water" that extinguished phlogiston.

The Reform of Nomenclature

It was not enough to demonstrate the composition of the four elements to kill that theory. Another concept of the element, defined as an undecomposable substance, the residue of analysis, had to be accepted. Lavoisier's definition is famous: "If we give to the name of elements or principles of bodies the idea of the last step that analysis can reach, all the substances that we have not been able to decompose by any way whatsoever are elements for us: not that we can be sure that these bodies, which we regard as simple, do not themselves consist of two or even more principles, but since these principles never separate, or rather, since we have no means of separating them, they act to us like simple bodies, and we must not suppose them to be compound until the time when experiment and observation will have furnished the proof" (Lavoisier, 1862, vol. 1, p. 13). Here again it must be recognized that this definition was not new, since people were tempted to attribute it to Boyle, and it was already in use in the eighteenth century. But the novelty was that Lavoisier presented it as an alternative to the definition of the element-principle as constituting bodies. He condemned the search for the ultimate constituents of matter as vain and metaphysical and proposed to construct a system of chemistry on the exclusive basis of this new, strictly functional definition, which made the element something relative and provisional.

This project was elaborated on the occasion of nomenclature reform (Crosland, 1962). Guyton de Morveau, chemist of Dijon—who was, as we have seen, very close to Bergman of Sweden, and who was charged with compiling the dictionaries of chemistry for the *Encyclopédie méthodique*—had undertaken the enormous task of reforming the nomenclature in 1782. The general idea was to indicate the composition of a substance by its name. In uncertain or controversial cases, Guyton, who was persuaded that a language is a question of convention, proposed an arbitrary or neutral name to obtain the agreement of his colleagues. Soon after the solemn experiment on water, he went to Paris to submit his project to the Academy. Then the little group of the "converted" persuaded Guyton to give up ownership of his project and revise the work as a team. So the new system was signed by four authors, Guyton de Morveau, Lavoisier, Berthollet, and Fourcroy, and was published in 1787 under the title of *Méthode de nomenclature chimique*. It consisted of an "alphabet" of thirty-three simple names for the simple substances: simple, familiar substances like copper and sulfur kept their old names, and those that had recently been discovered, particularly the "airs," were named after a characteristic property—for example, *oxy-gen* = "acid generator"; *hydro-gen* = "water generator"; *a-zote* (French for *nitrogen*) = "unfit for animal life." Compound substances were designated by a compound name juxtaposing the names of its constituents and classified according to genus and species; the generic name (for example, acid), designating the properties common to a whole class, was specified by an adjective—for example, carbonic acid. When two substances united to form several different compounds, they were distinguished by their suffixes: *-ic, -ous* for the acids (for example, sulfuric acid and sulfurous acid); *-ide, -ate* for the salts (for example, sulfide and sulfate); or by prefixes for oxides.

In comparison with Guyton de Morveau's earlier project, Lavoisier introduced two major changes. On the one hand, he based all the names exclusively on his theory, thus transforming the collective enterprise of nomenclature reform into a weapon against the doctrine of phlogiston. On the other hand, he changed the project's philosophy. For Guyton, nomenclature was a question of convention, and the agreement of all chemists was necessary for it to be valid. For Lavoisier, nomenclature had to reflect nature. This change took place during a long detour through the "metaphysics of languages" that Lavoisier borrowed from a contemporary philosopher, the Abbé Etienne Bonnot de Condillac (1715–1780). Language and knowledge were inseparable, so to reform the language was to remake the science. But where could the principles of a well-made language be found? In nature. Condillac's *La logique* taught that nature's method, analysis, "yeast for the mind," must be followed. Analysis—used in the double sense of a progression from the complex to the simple and from the simple to the complex—was the only way to avoid errors and prejudices. A language based on this "natural logic" was therefore much more than a lexicon, a "method of naming"; it was a program more than a finished product.

Two centuries later the principles of this nomenclature remain in use. Whatever their level of composition, compounds were always supposed to be binary. This general dualism, perhaps inspired by the chemistry of salts and displacements and brought up to date by electrochemical theory later on, in some way bridges the gap between eighteenth- and nineteenth-century chemistry—it is an element of continuity in spite of Lavoisier's break with tradition. But when it was published in 1787, the nomenclature was perceived rather as an offensive against phlogiston, and it set off a violent controversy. Twenty years later, in spite of reservations and criticisms about certain names—especially oxygen and *azote* (nitrogen)[32]—the new nomenclature was for all practical purposes adopted and taught in all European countries.

Lavoisier's Triumph

This victory was achieved through a campaign of persuasion led by Lavoisier and his collaborators: correspondence with chemists everywhere; invitations to dinner; the translation of Richard Kirwan's *Essay on Phlogiston* (with a formal rebuttal of his attempt at compromise); the creation of a new journal, the *Annales de chimie*, in 1789 to counteract the work of the two heads of the opposing camp (Jean-Claude La Métherie's *Observations sur la physique* and Crell's *Chemische Annalen*). Such was the triumph of the reformed nomenclature that fifty years later, in the mid-nineteenth century, pre-Lavoisier works on chemistry had become unreadable and were discarded into the dustbin of distant prehistory. This rupture also had a social dimension: the language of the chemists of the Academy was no longer the same as that of druggists and artisans, who would long continue to speak of "spirit of salt," "vitriol," and the like. Venel's efforts to defend the double language, both popular and scientific, were dead and buried. Chemistry was no longer the business of artists. In France, as in other countries where one distinguished between pure and applied chemistry, it was theory that inspired and controlled practice. Guyton, Fourcroy, and Berthollet, disciples of Lavoisier, would prove this with their brilliant technological and military exploits for revolutionary France.

Soon after the nomenclature reform Lavoisier published the *Traité élémentaire de chimie*. He presented it as a work that broke with traditional treatises. Addressing himself to "beginners," he attempted to construct chemical ideas from facts, proceeding according to "natural logic" from the simple to the complex. To demonstrate chemistry on the basis of analytic logic was to break its historic and pedagogical links with natural history. Traditionally, the two sciences were linked together and treatises were organized according to the three realms of nature: mineral, vegetable, and animal. That is why Lavoisier's treatise seems to be the crowning achievement of the chemical revolution and the first modern work of chemistry.

Lavoisier did not, however, reconstruct *all* of chemistry. Conscious of the limits of his treatise, he explained very clearly, in his "Discours preliminaire," the price he had paid to achieve his end. He named three conditions—that is, exclusions: he had had to leave out the question of the constituent elements of matter, to omit the history of chemistry and the work of his predecessors and contemporaries, and, most important, to avoid discussion of the chemistry of affinities. This theory was too complex for beginners, he said. But, in fact, the question of affinity could not be integrated into analytic logic easily. The relationships of displacement would remain strangers to the comings and goings of the simple and the complex, and Lavoisier resisted all hypotheses on the attractive and repulsive forces of simple bodies. In brief, his revolution left the traditional chemistry of affinities intact, untouched.

Moreover, Lavoisier had not really attained his goal. He was well aware of this, since he judged his work imperfect. Analytical order, from the simple to the complex, did not structure all of his famous *Treatise.* It was rigorously followed only in the second part, which presented substances in their order of composition in forty-seven tables. The first part showed the most recent results of pneumatic chemistry and antiphlogistic theory. And the third part described the operations and apparatus of the chemist, with illustrations by Marie-Anne Pierrette Lavoisier, his wife.

By stressing the limits of Lavoisier's work in this way, we do not seek to minimize or contest its impact. If we have chosen to present Lavoisier as the last figure of eighteenth-century chemistry rather than the first modern chemist, it was not with the object of taking the side of continuity in the debate that has, for two centuries, opposed "continuist" interpretations to celebrations of a break with the past. Our object was to show the nature and extent of Lavoisier's work, which was based on the tradition he received.[33] Lavoisier remained rooted in the eighteenth-century chemistry of salts and principles while, at the same time, he opened a new field of research by changing the objectives. In the *Treatise,* he tried to reorganize all of chemistry, both theory and practice, around analysis. He was no more able to complete this task than were his colleagues and contemporaries, Antoine François de Fourcroy (1755–

1809) and Jean-Antoine Chaptal (1756–1832). But he gave chemistry a logic of its own, which separated it from natural history, and he specified a program and a goal for it: "Chemistry's object in submitting different natural bodies to experiment is to decompose them . . . It therefore moves toward its goal and its perfection by dividing, subdividing, and resubdividing again, and we do not know when it will achieve success" (Lavoisier, 1862, vol. 1, pp. 136–137).

Lavoisier changed the practice of chemistry fundamentally. The object of his *Treatise* was to educate chemists in a year or two, to inculcate the foundations in them and train them in laboratory techniques. How could an apprenticeship in chemistry, which necessitated long experience, a familiarity, a "madman's passion," according to Venel, become a child's game, an affair of one or two years? It was the introduction of measurement that revolutionized laboratory work. The chemist in Lavoisier's school no longer needed a "thermometer in his fingertips" or the knack of an artist. He had thermometers, a calorimeter, gasometers, hydrometers, and, best of all, the precision balance. Yes, Lavoisier did revolutionize chemistry with the balance—not that it was unknown before him, but for Lavoisier it was not just a costly and sophisticated precision instrument. It was a key for deciphering nature. Playing with the famous idea of the conservation of matter—nothing is lost, nothing is gained—Lavoisier redefined the chemist as someone who put all chemical reactions into balance, whatever their complexity and diversity of circumstance. Weighing what went in and what came out of the reaction, making balance sheets for reactions, and putting them in the form of equations was the way to control the process of transformation, even if doing so enclosed the process in a "black box." The balance was therefore the preeminent instrument in the chemistry laboratory, an organizing concept for the abstraction of certain circumstances, and an instrument of argumentation that created a theater of proof—all at the same time. By offering the possibility of controlling phenomena, it also opened a path between science and technology. From the laboratory to the workshop, from pure theory to applications, the balance supplanted the alembic and the retort as the symbol of chemistry.

3

A SCIENCE OF PROFESSORS

Respected at Last

"Flemish gentleman of good family, wealthy and independent, with a taste for chemistry, is looking for . . . the absolute." It was a costly quest. Balthazar Claes spared no expense to equip his personal laboratory. He searched alone, day and night, tortured and possessed.

In *La Recherche de l'absolu* (1834), Balzac brought back to life an archaic figure, a kind of chemist that was disappearing. Although he studied in his youth with Lavoisier, Balthazar Claes was lured away from ordinary chemistry by a mysterious visitor and became involved in a demoniacal chemistry, a madman's passion that destroyed his character and ruined his family. On the fringes of worldly affairs, Balthazar appears as a strange mixture of the traditional portrait of the alchemist and the already obscure figure of the chemist of the "elements." In a period that valued material success, all the chemistries of the past were mixed up, all would be described pejoratively: "chemistry is a madman's trade," as Balzac complacently described the pit that swallowed up a fortune "under the smoking debris of silly fancies."

In counterpoint, *La Recherche de l'absolu* sketched the image of "normal science." Gabriel, the eldest son of Balthazar Claes, entered the Ecole Polytechnique. "He will be a scholar," said his father. And in fact Gabriel did represent the rising generation of scholars who made a career for themselves and prospered in elite schools and high offices. While the search for the absolute isolated the chemist, depleting fortunes and destroying families, the new chemistry was a social activity. It initiated new social and mercantile networks: learned societies and industrial firms. Great fortunes and brilliant careers were built on it.

A glance at the visiting cards of some French chemists of the nineteenth century shows the impressive array of career possibilities open to chemists. For example: Louis Joseph Gay-Lussac (1778–1850), professor at the Ecole Polytechnique, professor at the Jardin des Plantes, academician, a director at the mint, president of the Compagnie de Saint-Gobain, deputy for Haute Vienne, peer of France; Baron Louis Jacques Thénard (1777–1857), peer of France, counselor of public instruction, chevalier of the Legion of Honor, member of the Academy of Sciences, dean of the faculty of sciences of Paris, professor of the Royal College of France, professor of the Ecole Polytechnique, member of the Academy of Medicine, member of the Société Philomatique and of half a dozen academies in London, Berlin, and elsewhere.

The titles of Gay-Lussac and Thénard were not at all exceptional in the nineteenth century. Holding many offices was the rule; a seat in the senate and a ministry were not unusual perks in a renowned chemist's career. Each regime had its chemist. If Thénard owed his social prominence to the Restoration, Jean-Baptiste Dumas (1800–1884) flowered under the Second Empire: professor at the School of Medicine and the Sorbonne, lecturer at the Collège de France, he was first minister of agriculture and commerce (1849–1851), then minister of education at the beginning of the Second Empire, and he became permanent secretary of the Academy of Sciences in 1868.

Marcellin Berthelot (1827–1907) was the Third Republic's chemist: professor at the Collège de France and the School of Pharmacy, president of the Chemical Society of France, he was named inspector general of higher education in 1876, senator for life in 1881, minister of public instruction at the end of 1886, permanent secretary of the Academy of Sciences in 1889, and minister of foreign affairs in 1896. He was honored with a national funeral in 1907.[1]

Prestige, authority, and dignity had replaced the prejudice that had previously discredited the practice of chemistry. During the nineteenth century, chemistry was the very image of an exemplary science, a model of positivity. Renouncing the search for the absolute and vain inquiries into ultimate causes, and wisely limiting its terrain to the laws of phenomena, it progressed in giant steps. Alchemy had made promises. Chemistry was performing great feats! It seemed to give birth to innumerable applications spontaneously. It spread progress, comfort, and prosperity.

Before electricity, chemistry fed the literature on the marvels of science and industry. Its formidable power, which fostered dreams and terrors at the same time, inspired the figure of the modern Prometheus, Frankenstein. In her famous 1817 novel, Mary Shelley incarnated chemistry in the person of professor Walden, Frankenstein's master. She described chemists as modest scholars with dirty hands, bent over their microscope or retort, penetrating the secrets of nature to work miracles every day.

Master of a new world, symbol of prosperity and progress, the professor/chemist could aspire to the power of the alchemist without taking on the alchemist's dark and mysterious side. On the contrary—the chemist worked under the sun of reason. He symbolized the triumph of a science that, having abandoned its dreams of unlimited power, was building a monument to the glory of order and progress, discovery by discovery, on a limited but solid foundation.

Professionalization

The chemist had finally acquired public respect and recognition, and he had acquired it by unexpected means—the dynamic of institutions. A current but rather lazy interpretation makes the process of professionalization the ineluctable consequence of scientific progress. Sometimes the relationship is inverted and it is professionalization that gives progress its impetus. Chemistry in the nineteenth century bypassed both these routes. The creation of educational and research institutions of new types guaranteed the stature and the social status of the discipline; it furthered the spread of knowledge and the formation of an army of chemists. But this knowledge and army of chemists produced a real mobilization of natural phenomena by the scientific approach. If, in the global process of professionalization of the science, which affected all fields of knowledge in the first half of the nineteenth century, chemistry was often in the forefront, it was also because this process was nourished by a real transformation of chemical methods, and not just a formal one.

The movement was energized by the creation of specialized journals in various cities in Europe at the end of the eighteenth century. While assuring a rapid dissemination of information, the journals contributed to the development of links between chemists and the development of networks of specialists who would gradually weaken the larger community of the academies. In this respect the chemical revolution marked a turning point: Lavoisier presented his revolutionary work on the stage of the Royal Academy of Sciences, but, in order to spread the new doctrine, he and his disciples created their own organ of diffusion, the *Annales de chimie*, in 1789.[2] It was followed almost immediately by an Italian sister, the *Annali di quimica*, created in Pavia in 1790, and a Spanish sister, *Anales de química*, founded by Proust in Segovia in 1791. In Germany where the first chemistry journal, Crell's *Annalen*, had appeared, followed by *Algemeines Journal der Chemie* in 1798, it was the *Annalen der Pharmacie* created by Liebig in 1832 that assured the spread of chemistry.[3] In England the *Chemical Journal*, founded in 1798 by William Nicholson, soon had competition from the *Philosophical Magazine* and then the *Quarterly Journal*, the bulletin of the Chemical Society, which took the title of *Journal of the Chemical Society* in 1867.

The creation of chemical societies reinforced this process by knitting together national communities: the Chemical Society was founded in London in 1841, the Société de Chimie of Paris in 1857, the German Chemical Society in Berlin in 1866, the Russian Chemical Society in 1868, and the American Chemical Society in 1876. In each case a new journal, the official organ of the society, was created a year or two later. Moreover, chemistry played a pioneering role in the organization of the sciences, since it was the first discipline for which an international congress of specialists was organized (at Karlsruhe, in 1860).

But it was above all its rise in higher education that transformed the status of chemistry. In Spain, Germany, France, Great Britain, the United States . . . everywhere the teaching of the experimental sciences was developing, chairs in chemistry were multiplying. Chemistry took over little by little in a variety of fields—not only pharmacy and medicine, but also engineering and agriculture.[4] The phenomenon was international and the effects were massive. At the end of the eighteenth century, chemistry was cultivated in Europe by only a few dozen scholars, most of whom practiced other activities as well; toward the middle of the nineteenth century, chemistry was practiced full-time by hundreds of well-educated chemists.[5] Although this figure appears ridiculously small in comparison with the size of the student population in our own time, it nevertheless led to a change: chemistry was exercised as a profession, a full-time, paid activity for which one prepared with specific training and with studies sanctioned by diplomas.

Experimental work was recognized as a necessity in the chemist's education. It was no longer a question of making a few spectacular demonstrations to entertain a small audience, but of training pupils in laboratory work. After a pioneering attempt at laboratory education in Hungary, the Ecole Polytechnique initiated obligatory practical work in chemistry. Later Justus von Liebig (1803–1873) invented an original institution, the laboratory-school (Morrell, 1972; Fruton, 1990). A good chemist, Liebig said, is someone who knows how to see and to feel, to "think in terms of phenomena; who knows how to keep in his memory the sensations connected to the experiments and products he has manipulated in the past." In order to develop this faculty and acquire expertise in chemistry, a daily, intensive training in chemical manipulations under the guidance of a master was necessary. But the apprenticeship that produced that indispensable familiarity lasted only four years and not, as in Venel's day, a lifetime! Thus in a number of European cities schools of research, which presented characteristic styles and methods of investigation, developed around a professor-leader.

The big teaching laboratories needed laboratory assistants, whose tasks the directors preferred to entrust to young doctoral students rather than to amateurs. The flowering of chemistry in higher education thus encouraged professionalization in two ways, not only by the education offered but also by an increasing number of employment opportunities.

The conquest of academic territory by chemists took different routes in different countries. Nineteenth-century chemistry was both European and profoundly influenced by national styles. It was "European" because of the circulation of men and ideas from one country to another. Sending students abroad for a year or two seems to have occurred rather frequently and played a key role in the evolution of chemistry. But the practice also encouraged competition—between different forms of national institutional organization as much as different programs of research—that would create tension in nineteenth-century chemistry.

Organizations and Programs

In 1800, as a result of the two revolutions it had endured, chemical and French, Paris was the center of European chemistry. In spite of wars, French nomenclature was translated, distributed, and adapted all over Europe. And in France the war had encouraged the emergence of a new personage: the entrepreneur-chemist who saved the country. Guyton de Morveau, the first president of the Committee of Public Safety, and his colleague Claude-Antoine Prieur de la Côte d'Or pulled off a spectacular feat in April 1793 by creating a secret military laboratory to build observation balloons.[6] The Committee of Public Safety made another appeal to chemists to instruct their fellow citizens in the art of making saltpeter. Monge, Berthollet, Guyton de Morveau, and Chaptal, who was sent for from Montpellier, taught a revolutionary course on the making of saltpeter and canons. Finally, on the Committee of Public Instruction, Fourcroy supervised chemistry teaching. Chemistry enjoyed a prominent place in the ephemeral Normal School of Year III, at the Ecole Polytechnique, and in medical studies at the Ecoles de Santé. (From 1792 to 1805, the French followed a new calendar as part of the revolutionary regime.) Under the Consulate, Jean-Antoine Chaptal, as minister of the interior, devoted himself to the development of technical education. In 1800, in an *Essai sur le perfectionnement des arts chimiques*, he suggested that four colleges be created to teach the chemical trades.

In brief, the revolutionary period favored chemists. They were given more power, more credit, and more positions. They had lost Lavoisier, who was guillotined May 8, 1794, but the chemical community reformed itself in celebration of the memory of its "immortal founder" and prospered in the new institutions. The Ecole Polytechnique,[7] which admitted 120 students per year, was the pinnacle of scientific education throughout the nineteenth century. For young, ambitious scholars, the new rallying point was the Society of Arcueil, founded in 1802 by Laplace and Berthollet to develop a Newtonian program of physicochemical research (Crosland, 1971). It welcomed and promoted young graduates of the Polytechnique, like Dominique François Arago and Gay-Lussac.

France, however, would not long be a model for organization. A salient characteristic of nineteenth-century French science was its concentration in the capital. Toward the middle of the century Paris had more than half the chairs in chemistry.[8] The professors, overburdened with the work of their accumulated positions and the annual "forced labor" of the *baccalauréat* exams at the Faculty of Sciences, were even less inclined to do research because they had no funds to equip research laboratories, and because their advancement depended a great deal on the rhetoric in their lectures (Fox, 1973; Shinn, 1979). As a result, education in research took place outside the faculties, either in the higher schools or in the new institutes.[9]

The French provinces suffered greatly from the prestige of Paris, the center of attraction for students from all countries during the first half of the nineteenth century. There were only seven chairs scattered among the faculties of science of Strasbourg, Lyon, Montpellier, Toulouse, Nancy and Caen, to which three or four chairs in faculties of medicine or pharmacy can be added. Often the professors in the provinces felt sidelined, out in the wilderness. So Montpellier, which was a dynamic, respected university center and competitor of Paris in Venel's time, was seen as a place of exile around 1840 by the young and brilliant Charles Gerhardt, whom Dumas had taken away from Paris. In 1854, the minister of education directed a redistribution between Paris and the provinces, and the provincial universities became more dynamic and more innovative than Paris in some cases (Nye, 1986). A position in Paris was still, however, the only promotion worthy of a promising chemist like Louis Pasteur, who began his career at Strasbourg and then worked at Lille. And Pierre Duhem moldered away in Bordeaux until the end of his life because he opposed the thermochemical ideas of the all-powerful Marcellin Berthelot. In sum, centralization and the power of the mandarins went hand in hand.

The German case contrasts so sharply with the French one that it could be seen as a counter-model. Chemistry developed in pre-Bismarck Germany in a regional, decentralized framework. It was taught in traditional universities, but after the reform of the educational system inspired by Humboldt and Fichte, the university became a seat of culture and innovation. Liebig, who had studied in Paris with Gay-Lussac and Thénard, went back to the University of Giessen and opened a teaching laboratory. He attracted students from various countries to Giessen, and his teaching became a model for others in Germany and overseas.

Liebig's pedagogical model was particularly influential in the United States, where university chemistry education was organized by professors who had studied with Liebig or Friedrich Wöhler (1800–1882).

While Scotland had its own original research tradition—at Edinburgh, where Joseph Black taught until his death in 1799, and at the University of Glasgow, where Thomas Thomson gave practical and theoretical instruction that attracted both students and industrialists from 1818 to 1852—London followed the German model. The Royal College of Chemistry, founded in 1845, was directed by August Wilhelm Hofmann (1818–1892), a former assistant to Liebig, who had left Bonn for London (Bud and Roberts, 1984).

In France, on the eve of the Franco-Prussian War, the German model became a favorite reference in requests for funding for laboratories and research. Adolphe Wurtz, who had been a student at Giessen, addressed a report on "Higher Practical Studies in German Universities in 1869" to the minister of public instruction, Victor Duruy. He presented the creation of laboratories as a national investment: "capital invested for a high rate of return." Although in the heroic times of Scheele and Lavoisier a humble shop had been sufficient for great discoveries, argued Wurtz, modern chemistry required collective research in a modern, well-equipped laboratory: "A group of workers cluster around the master. They all profit from his teaching and his example, and each one from his neighbor's experience . . . A laboratory is therefore not only an incubator for science, it is a school." Joining the fight for funding, Louis Pasteur chose another metaphor: "Without their laboratories the physicist and the chemist are unarmed soldiers on the battlefield" (Pasteur, 1868). Judging from the report on chemistry laboratories written by Edmond Frémy (1814–1894), a professor at the Museum, these urgent appeals were heard, for he described a euphoric situation in 1881.

While the model of the laboratory-school spread widely, another characteristic of the German system was more difficult to transfer: the liaison between universities and industry (Fisher, 1978). It was a phenomenon of culture and an enormous asset, for the profession of chemist developed at the crossroads of the academic world and industrial production.

"Pure" Chemistry and "Applied" Chemistry

In the nineteenth century chemistry was no longer considered an auxiliary science of medicine, pharmacy, or geology but an end in itself. This change, sought by Venel when he wrote for the *Encyclopédie*, occurred in the first half of the nineteenth century within the categories of "pure" and "applied." Instruments of academic recognition in the eighteenth century, these categories now served the expansion of academic chemistry. Although the majority of chairs in chemistry were found in faculties of medicine or pharmacy and schools of agriculture or mining, a general course of theoretical chemistry was usually taught. Applied chemistry implied pure chemistry. That is why the promotion of chemistry as a useful science in the service of industry, agriculture, or health benefited the community of academic chemists as much as, if not more than, entrepreneurs.

Moreover, to have new courses approved, the professors used the argument of utility, before any industrial demand for skilled chemists had arisen. We don't know of any cases of industrialists applying pressure to have university teaching focused on industrial education (Donnelly, 1991). Applications were usually an excuse, for once a course in applied chemistry was under way, its original purpose was often forgotten for the sake of academic chemistry. For example, England's Chemical Society, created in 1861, was intended to be a means both for the advancement of science and for improving industrial practice. The purpose of the scientist was "the general advancement of Chemical science as intimately connected with the prosperity of the manufactures of the United Kingdom, many of which depend on the application of chemical principles and discoveries" (Bud and Roberts, 1984, p. 69). Soon, however, the publication of research became the chief purpose of the society, which was increasingly dominated by academics.

This professional strategy implied a view of technology as a simple application of pure science, a process deprived of any specific creativity. At the same time there were some protests against the utilitarian bias that tended to prevail in scientific research. The French philosopher Auguste Comte considered a clear-cut separation between scientific research and technological demands a criterion of civilization (Comte, 1830). Liebig vehemently denounced utilitarianism in science and advocated a disinterested pursuit of knowledge. This kind of rhetoric proved remarkably effective (Liebig, 1863). The chemistry of the "artists" gradually gave way to a professional chemistry based on academic curricula. Qualified chemists were actively involved in the transformation of chemical industry rather than in fulfilling preexisting industrial needs. In some cases, industrial demand followed the availability of workers.

Such was the miracle accomplished by the chemistry professors of the nineteenth century.

Did that mean that chemical technologies were actually submitted to the standards of the theoretical system; that "industry is the daughter of science," as some chemists said? This fundamental question will underlie our presentation of nineteenth-century chemistry.

Throughout the nineteenth century, chemistry transformed the countryside, clothes, health, and daily life. Extractive materials were gradually supplanted by artificial products. This process, which is still going on, began in the nineteenth century at the cost of a few tragedies and quick conquests.

Since the ascent of academic chemistry was matched by that of industrial chemistry, are we going to describe the scientific and industrial developments jointly or dissociate them? In a number of cases a close relationship between the two is obvious. One cannot understand the change from the hand-work of artisans to the mass production of standardized products without taking account of the methods of analysis, the batteries of tests, and the protocols of experiments defined in the laboratory. One cannot imagine the synthesis of dyes or medicines without the formulas and reagents developed in research labs. On the other hand, technical perspectives and innovations stimulated research programs and specialized training. At the beginning of the century a chemist and a mechanical engineer working together could transform a laboratory reaction into an industrial procedure, but by the 1880s the limits of the model of applied science faithfully following the progress of academic science were felt. The demands for special qualifications gave rise to what became known as chemical engineering. Initiatives were taken to make appropriate instruction available to educate chemical engineers in various countries: by George Davis in Manchester in 1880; by the Alsacian chemist Charles Lauth and Paul Schützenberger (1829–1897), who created in 1882 a School of Industrial Physics and Chemistry in Paris.[10] Especially in the United States, with programs at the Massachusetts Institute of Technology in 1888, the University of Pennsylvania in 1892, and Michigan in 1898, chemical engineering developed. The program was organized around the concept of "unit operation" introduced by Arthur D. Little. The complex mechanisms of a production process were broken up into smaller units or sequences that were studied separately, transposed, and reused in a variety of processes (Guédon, 1983),

making it possible to base chemical production on the model of mechanical production.

In addition to chemical engineering, another profession whose objective was to rationalize invention emerged at the end of the nineteenth
century: industrial research. In France, the chemist Henry Le Chatelier
(1850–1936), a member of the Society for the Encouragement of National
Industry, recommended the industrialization of science on the model of
the Taylor system; in Germany, industrial research took place in laboratories established within the company structure of industrial firms. Thus
began the kind of applied research in which invention was no longer given
over to the caprices of chance or genius but was carefully programmed.

More generally, although professors of chemistry and chemist-entrepreneurs were mutually supportive, the history of industrial chemistry
cannot be reduced to a history of applied chemistry. The development
and success of a procedure or a material depended on a series of technological, economic, and political constraints—the integration of procedures, the laws of the marketplace, the organization of technical
education, legislation and patents, national tensions and wars—as much
as on scientific knowledge. Industrial chemistry requires the control of
reactions, but also that of production, invention, and, last but not least,
markets. It emerged in an atmosphere of war, conquest, and often exac-
erbated rivalry: the blockades, customs barriers, and cartels that operated while each country struggled to dominate the market. At the
beginning of the nineteenth century, France was in the forefront of
chemical industry; then Great Britain took the lead; in the last quarter of
the century, Germany achieved supremacy, and after the First World
War the United States made a spectacular ascent.

It is therefore necessary to stand aloof from the categories "pure" and
"applied" that dominated the rhetoric of nineteenth-century chemists.
In dealing separately with scientific and industrial chemistry in two
successive sections, we have chosen to emphasize the fact that industrial
development did not take place entirely under the tutelage of the
scientific establishment; there was, in fact, a host of other factors—capital, patents, wars—that escaped the realm of the chemistry professor. In
spite of its power, the duo of pure and applied chemistry could not rule
the world of production.

The separation of industrial and academic chemistry also allows us to
present scientific chemistry as an original enterprise organized around

education. The rising power of the professors marked the evolution of chemical doctrines profoundly, because pedagogical preoccupations influenced research and techniques. An independent scientist with time and a laboratory of his own practices chemistry in ways that differ greatly from those of a professor overloaded with classes, confronted with new students every year, and in charge of organizing collaborative research. Although the time spent on teaching duties can slow the dynamism of research, the need to organize knowledge in order to transmit it and to train qualified chemists can encourage discoveries. The pedagogical challenges that had inspired Lavoisier in his *Treatise* were faced repeatedly in the nineteenth century. In their attempts to write logical, clear, and systematic textbooks, professors had to answer scientific challenges, such as the classification of simple and compound substances.

Academic textbooks therefore provide access to the chemistry of the nineteenth century. They have both a normalizing and a mobilizing function. On the one hand, they can be viewed as a product of research, a specific way of recording results that standardized information and organized it in order to facilitate understanding and memorization. In this way textbooks made it possible to define the discipline as a body of organized, standardized, and teachable knowledge. That is why the science presented in textbooks is often depicted as frozen or dead: the tortuous paths of research have been removed; battles with experiments or colleagues go unmentioned. But this work was useful precisely because it set a proper dynamic for the discipline. In their pedagogical finality, the textbooks were also meant to enroll new recruits, to prepare the armies of chemists who would eventually occupy positions in the academy or industry. In this way they played a role of primary importance in the process of professionalization, which resulted in the modern figure of the chemist. As Primo Levi pointed out (and as mentioned in the Prologue), the textbook, with its areas of shadow and light, defined the world in which the destiny of the future professional chemist was sealed. Through academic textbooks—occasionally supplemented by articles and correspondence—we will try to identify the main themes these professors promulgated and used to increase the potential and the population of chemists. We will call them "programs," in the double sense of collective research projects that defined a "school of research" and curriculum plans that set the standards of knowledge necessary for professional practice.

Broadly speaking, three aspects of chemistry can be identified, three successive but not exclusive programs. Each was organized around a specific intellectual and experimental practice: decomposition, substitution, and synthesis.

At the beginning of the nineteenth century, analysis, which Lavoisier had defined as the goal of chemistry, was the essential objective that all innovations—experimental techniques, theories, concepts, laws—were designed to serve. The chief result was the growing number of identified simple bodies. This raised the problem of managing them all, and of managing the classification of substances and information as well.

Beginning in the 1840s, substitution became the organizing concept of a new field, organic chemistry, which in turn transformed inorganic chemistry, doubled the management problems, and refocused efforts at classification. Around 1860, synthesis developed as a new theoretical and practical program, with new concepts, new disputes, and new problems involving formalization and optimization.

The traditional image of a positive science, solidly anchored on Lavoisier's foundation, advancing in giant steps and then giving birth to a host of applications, will presumably be altered by our narrative. Nineteenth-century chemistry will be presented as a less linear, more confused process of development. Of course each "program" threw light on and surpassed the preceeding one, but it did not resolve all the problems— and sometimes it revived them. Each step, each advance in theory or in experimental technique helped to redefine the identity of chemistry.

Analysis, a Galvanizing Program

"Chemistry moves toward its goal and toward its perfection by dividing, subdividing, and resubdividing again." This program, formulated by Lavoisier in 1789 in his *Treatise*, was more relevant than ever at the beginning of the nineteenth century. It was encouraged by two new discoveries, one instrumental (the electric cell) and the other conceptual (the atom). Attempts to isolate, identify, name, classify substances—activities that were the "hallmark" of scientific chemistry—would henceforth have the simple body as their primary focus and the precision balance as their preferred instrument. What is more, the balance was no longer just a means, it directed the chemist's program: to define each simple body by its weight—that is, by the measurable quantity entering into combinations. This was the objective, the obsession that motivated and reorganized the discipline. The chemical reaction was no longer an object of study in itself but a means of determining the elementary composition of the products of reaction.

Analysis developed both as a series of laboratory techniques for monitoring production and as an investigational program that built chemical theory along the axis from simple to complex. Inextricably theoretical and practical, analysis was a field in which, between 1800 and about 1840, cognitive and commercial interests were allied, in which the articulation between pure and applied chemistry took shape, in which the profession of chemist was born.

Fine Controls

If Lavoisier redefined chemistry with analysis as its center, his colleagues and successors perfected and diversified the *practices* of analysis (Svabadvary, 1966; Roth, 1988). Titrating solutions, assaying coinage, and detecting frauds were methods of control indispensable to the rise of the chemical industries, which needed definite, quantitative procedures. Fourcroy expanded the use of gravimetric analysis with hydrogen sulfide to detect traces of lead in wine or clarified cider. The burette, invented by Decroizille in 1795 and called the Berthollimeter, measured the quantities of reagents in a solution. In order to develop the bleaching industry, Berthollet began volumetric analysis to measure the amount of chlorine in "eau de Javel" (potassium hypochlorite manufactured in England as bleaching powder) and worked out simple testing procedures that could be handled by a workman without recourse to a specialist. For the analysis of minerals, and more especially for testing the gold or silver content of a metal, the *Manuel complet de l'essayeur,* published by Nicolas Vauquelin in 1799 and republished until 1836, outlined wet and dry methods that are still in use today at the mint. Gay-Lussac, who invented the wet assay method, was the titration champion in various areas: chlorometry, alkalimetry, sulfurometry (measurement of sodas, potash, bleaching powders, respectively).

Gay-Lussac provides a good example of how the articulation between pure and applied chemistry worked in analysis. A former pupil of Berthollet at the Polytechnique and a member of the Society of Arcueil,

Gay-Lussac accumulated teaching jobs and administrative and industrial responsibilities. Crosland (1978) presented him as a representative example of the first generation of professional, full-time chemists with large incomes. His publications reflect the diversity of his institutional affiliations: more than 150 articles published in the *Memoirs de la Société d'Arcueil* and the *Annales,* a hundred *Instructions* for entrepreneurs, and important technical innovations. Among his activities at the Compagnie de Saint-Gobain, Gay-Lussac proposed various titrating techniques that proved very useful in guaranteeing a uniform quality for Saint-Gobain's products. Moreover, he introduced an important improvement, a tower that increased the profit in the production of sulfuric acid, an essential ingredient in chemical manufacturing. Gay-Lussac practiced one and the same chemistry in the laboratory and in the factory. His objective was analysis, measurement, fine control. His method was precise and detailed protocols for experiments, which guaranteed the reliability of his analyses. The results were spectacular. Gay-Lussac contributed as much to the advancement of chemical science with his famous law on gas volumes and his studies of iodine and cyanogen as he did to industrial prosperity. Such a feat, however, implied limits in both fields of operation: Gay-Lussac's chemistry was purely experimental—indifferent to theoretical debates. And the alliance between pure and applied chemistry remained an individual achievement, which did not lead to true cooperation between science and industry.

Among Gay-Lussac's students, however, there was one who, while remaining in the academic world himself, managed to create the conditions for true cooperation between the world of knowledge and that of production. Justus von Liebig, professor at Giessen, widened the field of analysis to include organic compounds. He improved the laboratory equipment, introducing a combustion apparatus that was a crucial technical innovation in the analysis of non-nitrogenous compounds. He simplified analytic procedures, not only performing analyses more quickly than his predecessors but reducing the level of skill formerly required. Liebig thus transformed delicate analyses previously reserved for senior experimenters into easy operations that could be completed routinely by the students he was educating at Giessen. Liebig had created an original pedagogical system, a laboratory-school in which the students were asked to practice qualitative and quantitative analysis six hours per day, six days per week. While training his students by means of daily experimental practice, Liebig was able to include their results in his own research publications and to shape a "research school" (Fruton, 1990, chap. 2). Liebig can be portrayed as a "chemist breeder" (Morrell, 1972). From 1824 to 1851 the laboratory at Giessen was the best place for "breeding" a new style of chemist, expert in all the techniques of analysis and in the preparation of pure products—in other words, in the standard operations that allowed the changeover from artisanal fabrication to industrial production. Liebig's students learned chemistry as a functional discipline based on analysis. Undoubtedly, they could no longer speak Venel's "double language," nor understand the twists and turns of artisanal techniques, but they were perfectly prepared to supervise industrial production.

Liebig's pedagogical system had a tremendous impact on the evolution of chemistry. Among the hundreds of students who spent time in the Giessen laboratory were many future professors—Wöhler, Bunsen, Kekulé, Wurtz, Regnault, Williamson, Playfair—as well as captains of industry and entrepreneurs. The routinization and standardization of analytical practices led to a diversification of professional choices for academic chemists and brought well-trained chemists, who would be available to launch new industries, into the manpower market.

Volta's "Pile"

If analysis became a routine exercise, it did not develop in a peaceful setting devoid of "speculative" controversies. Quite to the contrary, the discovery of the electric pile (battery) raised the question of the effect of electric current on chemical compounds and revitalized the desire to reach a deeper understanding of nature through chemistry.

Gaston Bachelard considered a scientific instrument as "materialized theory," putting theoretical principles to work in a laboratory setting. In our contemporary view, the pile exemplified Bachelard's epistemology because it indeed materialized electrochemical theory. This was not, however, the original purpose of the pile made by Alessandro Volta (1745–1827), a professor at Como, and presented in 1800 to the Royal Society in London in a note written in French entitled "On the electricity excited by simple contact with different types of conductive substances."[11] Even if the cell was a concrete/abstract object from the beginning, it did not embody an electrochemical theory. Designed as an artificial electric organ to mimic those of electric fish, it was meant to refute the hypothesis of an animal electricity generated in the nerves, which had been advanced by Luigi Galvani as a result of his experiments on the contraction of frog muscle.

Volta was a physicist educated in the tradition of Laplace. To prove that there was only one kind of electricity, that of the physicists, he replaced the muscle and nerve of Galvani's frog with bits of cardboard and wet rags stretched between iron and copper. This analog of Galvani's experiment was presented as a column in which copper disks alternated with tin disks (later, silver disks with zinc disks). Each pair of disks was separated by small pieces of damp cardboard or leather soaked in water or another liquid.

This "pile," or sandwich, generated electricity, as a Leyden jar does. From this experiment Volta concluded that the electricity was due only to the contact of the metals. The conflict between Galvani and Volta was arbitrated by the Royal Society of London in 1800. Although we still use the term *galvanic electricity* and the metaphoric verb "to galvanize," Galvani was the loser. His career was ruined, and electrophysiology would have to wait several decades for recognition. Volta was celebrated as a hero. Invited to Paris in 1801, he presented his invention at the Institut, and Bonaparte, after giving him a medal, ordered an enormous battery of 600 elements for the Ecole Polytechnique's chemistry laboratory. The voltaic pile captured the interest of chemists, who took it as their own and completely forgot the debate over physiology. The invention of a physicist pirating an observation from physiology therefore opened a new career to chemistry.

The famous British chemist Sir Humphry Davy (1778–1829) later remarked that Volta's pile acted like "an alarm bell among the experimenters of Europe" (Knight, 1992, p. 39). Why so much excitement among chemists? As early as 1800 William Nicholson (1735–1815) and Sir Anthony Carlisle (1768–1840) used the cell to develop a new technique for analysis, called "electrolysis," that decomposed substances that had previously resisted heat and chemical action. But the first experiments in the decomposition of water using a voltaic pile resulted in confusion. The release of hydrogen at the negative pole and oxygen at the positive pole could be viewed as a confirmation of Lavoisier's views on the composition of water. But the subsequent formation of an acid and a base remained to be explained. So there was a new instrument to use for analysis and a new area for investigation as well (Knight, 1992, pp. 59–72).

What was the source of the current produced by the battery? Volta thought it was an effect of the contact between metals. This opinion, developed by Davy, was opposed by Michael Faraday (1791–1867), who proposed a chemical interpretation of the battery and introduced the terms *anode* and *cathode*. Faraday did experiments on the electrolysis of water and of several hydracids before he formulated a first statement, in 1823, of what later came to be known as the first law of electrolysis: "The chemical power of electricity, like the magnetic force, is in direct proportion to the quantity of the electricity which passes." Faraday designed an instrument, the volta-electrometer, later renamed "voltameter," to measure a quantity of electricity by means of the quantity of hydrogen produced by it. This apparatus illustrates the complex interactions between instrumentation and theory, since it is based on the very hypothesis of a relationship between the degree of chemical affinity of two elements and their ease of moving to opposite poles in electrolysis, which it then helped to test. In 1834, having passed the same amount of current through several liquids—water, tin chloride, lead borate, hydrochloric acid—Faraday developed the notion of the "electrochemical equivalent," corresponding to the weight of various substances decomposed by the same amount of electricity. So electrochemistry was sketched out. It was only after the development of this new area of research that the battery could become an objectified theory and the constantly perfected experimental device in use today.

Electrochemical Dualism

This new experimental tool was answered by a new theory, which explained the action of the battery with the help of a new model of the compounds it decomposed. On the basis of his experiments in electrolytic decomposition, Jöns Jacob Berzelius (1779–1848) of Sweden developed an electrochemical theory of combination around 1810. He defined each simple substance and each compound body by a positive or negative polarity whose intensity varied according to the nature and number of the positive and negative charges carried by the atoms of each element. The voltaic battery was no longer a simple instrument but a principle of intelligibility: electricity was seen as the primary cause of all chemical action, the "electrical fulfillment" of the Newtonian dream.

The intensity of the positive or negative electric charge indeed determined the combining force of elements, their degree of affinity. On the basis of two opposite forces alone it was possible to design a simple instrument for predicting chemical reactions: a scale of elements from the most electropositive to the most electronegative, from potassium to oxygen. Hydrogen was in the middle, between the electropositive and the electronegative groups.

Electricity thus seemed to solve the mystery of affinity that Lavoisier had not dared approach and that had obscured the controversy between Berthollet and Proust. Thanks to the electrochemical theory, affinity was now well integrated into the framework of a dualist chemistry of combination. Berzelius even revitalized Lavoisian dualism at the time it was beginning to totter under Davy's attacks. As early as 1810, when he established the simple nature of chlorine, which had until then been defined as "oxygenated muriatic acid,"[12] Davy had raised some doubts on Lavoisier's theory of acids and salts. Lavoisier admitted only oxyacids. Davy showed that there were also hydracids, consisting of a radical and hydrogen, which could be replaced by a metal, for example, to form a salt. The idea of substitution that would inspire the unitary theory of the future was not far off.

But as long as Berzelius, with his great authority and his international reputation, reigned supreme over European chemistry, at least until 1840, a salt would always be defined as the union of two bodies, simple or compound, of opposite electric charge. Salts would be the favorite topic of chemistry professors, the essential subject of their teaching. And chemistry would always be learned by proceeding from the simple to the complex, by studying the different combinations according to the order of their composition:

- the first-order compounds were formed directly by the union of two simple bodies—thus the combination of oxygen with a metal or nonmetal formed acids or basic oxides
- the second-order compounds resulted from the union of a basic oxide and an acid oxide to form a neutral salt
- the third-order compounds were the double salts resulting from the combination of two neutral salts
- finally, the fourth-order compounds were formed by the addition of water to a double salt

Berzelius's electrochemical dualism rescued Lavoisier's organizing principle for a time. It provided chemistry with a broad, theoretical framework that proved extremely relevant for the analysis of mineral substances.[13]

As Black and Lavoisier had integrated the study of heat into chemistry, Davy, Berzelius, and Faraday (who was perceived as a chemist by his contemporaries) added the study of electricity to chemistry's realm.

A "Demographic Explosion" of Simple Bodies

The annexation of electricity by chemistry created a major challenge for the chemist-professors, because the use of electrolysis led to a tremendous increase in the number of identified simple bodies. As soon as he understood the power of the voltaic pile, Davy used it to decompose substances that had previously been resistant to analysis. In 1807 and 1808 he successively isolated sodium, potassium, strontium, boron, calcium, magnesium, and, around 1810, chlorine, iodine, and bromine. Berzelius kept up the competition: from his laboratory came cerium, selenium, silicon, zirconium, thorium, lithium, and vanadium, and at the same time a host of new methods of quantitative and qualitative analysis. In short, the number of simple bodies nearly doubled. The table Lavoisier made in 1789 contained 33 simple substances. In 1834 Thénard named 54 simple bodies in his textbook, and in 1869 Dmitri Mendeleev counted 70.

The rapid inflation in the number of simple bodies marked nineteenth-century chemistry profoundly. The dominant impression of those involved was of fecundity. In fact, the program of analysis defined by Lavoisier had worked so well that chemists were faced with the necessity of admitting that the number of simple bodies was indefinite and undeterminable. Professors of chemistry found themselves in the greatest perplexity. They were forced to recite for their students the properties of an interminable list of simple bodies and to review the various combinations of each element. Was chemistry about to become a mere collection of monographs on substances, once again like natural history, from which the logic of analysis had distanced it? Would the clear pattern of a system be reduced to a simple inventory of individual properties that had to be memorized without understanding much about them? The textbooks of the first half of the century adopted the general division between metalloids and metals. This was acceptable from the theoretical point of view but inconvenient from the pedagogical one, because the list of metals was long and their compounds were numerous and particularly important for industrial applications.

A muted tension between two opposing tendencies underlies most of the nineteenth-century textbooks. On the one hand, the practical tendency of analytic chemistry inclined the authors to immerse themselves in the study of the particular, to treat chemical substances as individuals. On the other was a desire to generate and formulate general ideas and rules on chemical combination, on the composition of bodies. Far from diminishing, this tension grew stronger as chemistry became professionalized. The professional chemist, who worked in academic research or industry, sometimes devoted his whole life to a single substance, simple or compound, and that one substance determined the success or failure of his career. This is the destiny that the chemist-novelist Primo Levi deciphered, long afterward, from the pages of his student text. The professor of chemistry had to transform this assemblage of individual discoveries into a body of doctrine, a teachable discipline.

Analysis vs. Atoms

To respond to the challenges of experimental chemistry, which was accumulating new materials at a rapid rate, a theory was needed to discipline the population explosion of substances. The development of a theory was a pedagogical necessity as much as an epistemological precept. The professors' answer was the positivist concept of the scientific theory as a simple ordering of phenomena, an aid to memory. They believed theories and hypotheses to be precious, indispensable instruments—but on the condition we not put too much faith in them.[14] This theoretical strategy did, however, pave the way for gaining access to an invisible reality beyond the visible phenomena.

Dalton's Hypothesis

In 1804 a professor at Manchester, John Dalton (1766–1844), formulated a hypothesis that identified chemical elements with atoms. This notion, invented in antiquity on the shores of the Mediterranean and exploited in every possible way by the natural philosophers of the seventeenth and eighteenth centuries, was clearly not new. As Arnold Thackray pointed out, no doubt it was necessary to be, like Dalton, mostly unaware of the complexities of Newtonian chemistry to dare reactivate in an almost naive way an idea so laden with history (Thackray, 1966, 1970; Cardwell, 1968). But Dalton's atom was not a descendant of the ancient atoms or the Newtonian corpuscles. It was invented and put to work in another context.

At first Dalton was interested in meteorology and the physical properties of gases, and it was probably to explain the differences in the solubility of gases that he began to use the notion of the weight and size of atoms. Starting, as Lavoisier did, with the idea that gases consist of corpuscles that repel each other when exposed to heat, Dalton seems to have arrived at the conclusion that it was necessary to differentiate the corpuscles or atoms of the gases not only by their size or shape but also by their weight. How to determine the relative weight of atoms? Dalton confronted this question by turning to the quantitative or stoichiometric chemistry developed by Karl Friedrich Wenzel and Jeremias Richter at the end of the eighteenth century in Germany.

As we saw in Chapter 2, Wenzel and Richter were trying to define the relative affinities of bodies from equivalence tables of acids and bases. The "equivalent" quantity of a base is the weight of it necessary to neutralize a given quantity of acid, and vice versa for the equivalent quantity of an acid. Stoichiometry took on a different meaning in 1802, when, after a number of experiments on tin, antimony, and iron, Joseph Proust formulated a general law: "The relationships of the masses according to which two or several elements combine are fixed and not susceptible to continuous variation." This law of "definite proportions" extended the notion of equivalence, until then reserved for neutralization reactions between acids and bases, to all combinations. While Berthollet was involved in a long dispute over definite proportions in chemical combinations, John Dalton made Proust's law the basis for a new atomic hypothesis. He suggested that chemical combination takes place in discrete units, atom by atom, and that the atoms of each element are identical. He added a law of multiple proportions to Proust's law: when two elements form more than one compound, the various weights in which one element will combine with a fixed weight of the other (to form different compounds) are in a simple numerical ratio. Without the atomic hypothesis, added Dalton, these laws would be as mysterious as Kepler's laws before Newton. With the atomic hypothesis and the symbols Dalton used to summarize his system, they acquired an immediate, intuitive meaning.

Dalton's hypothesis made the notion of proportion—combination by discrete units—indisputable, but it shifted the debate to another problem: what was the correct formula? While equivalents deal with the ratios between simple constituent bodies, atoms require a number: how many atoms in a certain compound? How can the exact proportion of hydrogen that unites with oxygen to make water be determined? Like Richter, Dalton trusted the gravimetric ratios of combination and adopted a simple rule: when two elements form a single compound, it is binary and combines an atom of the one with an atom of the other. When they form two compounds, one is binary (combining an atom of each species), the other is tertiary (combining two atoms of one with one atom of the other), and so forth (Dalton, 1808). Water was thus described as a binary compound of hydrogen and oxygen with the relative weights of the two atoms being approximately one and seven. Ammonia was a binary compound of hydrogen and nitrogen whose relative weights were approximately 1 and 5. With this rule, called the principle of simplicity, Dalton could calculate the numerical values that allowed him to set the relative weights of atoms from his experimental data.

Unlike Richter, who could determine the equivalent by using the neutrality of the compound, Dalton had to relate all the atomic weights to a unit fixed in the conventional way. He chose hydrogen for the unit: the atomic weight of each element was the gravimetric proportion that combined with a gram of hydrogen to form the most stable combination.

This was the atomic hypothesis in its earliest version (Rocke, 1984). The Daltonian atoms had only a distant resemblance to their ancient homonyms. Their definition was different: they were no longer defined as minimum units of composition but as minimum units of combination. Their function was different as well: it was less a question of explaining a complicated visible object by an invisible simple one, as Jean Perrin would later say, than of resolving problems of language, formulas, and classification. Dalton's atoms differed from the Newtonian corpuscles just as much, since they presupposed neither the void nor attraction and made no attempt to explain the properties of simple bodies in terms of a complex architecture whose ultimate constituents would be atoms.

But atomic weight was a seductive concept. Instead of determining the composition of a body by percentages, a chemist could express it in terms of constituent atoms, establishing a direct link between experimental fact and its interpretation. Atomic weight was just what chemists needed for the task that confronted them: to characterize, name, document, and classify a continuously growing population of simple and compound substances. One could predict a great future for it.

Was Dalton's hypothesis well received? Before Dalton published his *New System of Chemical Philosophy,* Thomas Thomson, a professor at Glasgow, made himself the propagandist for the atomic hypothesis and gave its first public account in the third edition of his *System of Chemistry.* The atom was introduced in France by the translation of Thomson's *System* in 1809. In a long preface to the translation Berthollet declared that Dalton's principle of simplicity was arbitrary and warned against its seductions. The marriage of English ideas and French chemistry had a stormy beginning.

On December 31, 1808, just after the publication of Dalton's *New System of Chemical Philosophy,* Gay-Lussac announced that "volumes of gas that combine with each other were in direct proportion and the volume of the combination thus formed was also in direct proportion to the sum of the volumes of the constituent gases." In our view, the results obtained on the two sides of the English Channel converge harmoniously. The volumetric proportions seemed to confirm the gravimetric ratios. But Gay-Lussac rejected Dalton's hypothesis. And reciprocally Dalton contested Gay-Lussac's law. It implied that the number of atoms in a given volume of gas was the same in all cases, whereas according to Dalton's hypothesis, the formation of nitrogen oxide, which occurred at constant volume, necessitated that the number of atoms of oxygen and nitrogen per unit volume before the union be half that of the combined atoms of hydrogen and nitrogen for the same volume. This difficulty, which was insurmountable before the distinction between the atom and the molecule of a simple body had been made, set off a misunderstanding of several years' duration. Berzelius swept it away around 1819 when he used Dalton's atomic hypothesis and Gay-Lussac's law of volumes conjointly to determine a new system of atomic weights.

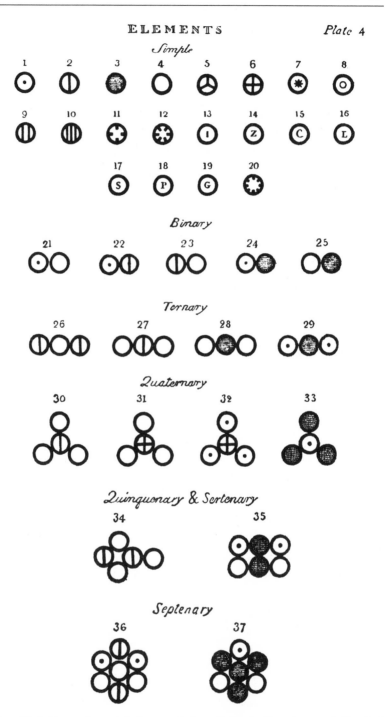

Plate IV (with caption on facing page) from Dalton's *New System of Chemical Philosophy* (1808)

PLATE IV. This plate contains the arbitrary marks or signs chosen to represent the several chemical elements or ultimate particles.

Fig.			Fig.		
1	Hydrog.; its rel. weight	1	11	Strontites	46
2	Azote	5	12	Barytes	68
3	Carbon or charcoal	5	13	Iron	38
4	Oxygen	7	14	Zinc	56
5	Phosphorus	9	15	Copper	56
6	Sulphur	13	16	Lead	95
7	Magnesia	20	17	Silver	100
8	Lime	23	18	Platina	100
9	Soda	28	19	Gold	140
10	Potash	42	20	Mercury	167

21. An atom of water or steam, composed of 1 of oxygen and 1 of hydrogen, retained in physical contact by a strong affinity, and supposed to be surrounded by a common atmosphere of heat; its relative weight = 8
22. An atom of ammonia, composed of 1 of azote and 1 of hydrogen 6
23. An atom of nitrous gas, composed of 1 of azote and 1 of oxygen 12
24. An atom of olefiant gas, composed of 1 of carbone and 1 of hydrogen 6
25. An atom of carbonic oxide composed of 1 of carbone and 1 of oxygen 12
26. An atom of nitrous oxide, 2 azote + 1 oxygen 17
27. An atom of nitric acid, 1 azote + 2 oxygen 19
28. An atom of carbonic acid, 1 carbone + 2 oxygen 19
29. An atom of carburetted hydrogen, 1 carbone + 2 hydrogen 7
30. An atom of oxynitric acid, 1 azote + 3 oxygen 26
31. An atom of sulphuric acid, 1 sulphur + 3 oxygen 34
32. An atom of sulphuretted hydrogen, 1 sulphur + 3 hydrogen 16
33. An atom of alcohol, 3 carbone + 1 hydrogen 16
34. An atom of nitrous acid, 1 nitric acid + 1 nitrous gas 31
35. An atom of acetous acid, 2 carbone + 2 water 26
36. An atom of nitrate of ammonia, 1 nitric acid + 1 ammonia + 1 water 33
37. An atom of sugar, 1 alcohol + 1 carbonic acid 35

But even in England Dalton was not unanimously accepted. Some chemists, who had been quick to recognize the immense usefulness of the concept of atomic weight, saw no reason to accept the atomic hypothesis. Why venture into an area inaccessible to experiment to establish the gravimetric proportions of combination? Davy continued to use the expression "proportional weight." William H. Wollaston, who was one of Dalton's first disciples, chose to substitute the expression "equivalent weight" for "atomic weight." He preferred to determine all these equivalent weights in relationship to the basic unit O = 100. It was only a question of convention and conversion was not very difficult. One simply multiplied by 1.25 all the values for the base H = 1. *Equivalent* and *atomic* were still almost synonymous. And *equivalent* was sufficient for the needs of analytical chemistry. The atom was superfluous.

The situation was actually extraordinarily complex, because the Daltonian atom did not eliminate the corpuscularism of the eighteenth century with the stroke of a magic wand. If chemists refused to reach a decision on the existence of chemical atoms, they mostly admitted the atom to the rank of basic notions that one defines in the introduction to a treatise, nevertheless. The Newtonian consensus on the corpuscular structure of matter persisted.[15] Thénard, for example, did not doubt that matter consisted of particles, whether they were called corpuscles, atoms, or molecules. If one could determine their weight, their size, and the force necessary to combine them, one would be able to deduce all chemical phenomena by means of geometry. Buffon's Newtonian views remained unquestioned, as if they were an axiom of physics and chemistry (Thénard, 1816, vol. 1). But the atom itself was beyond the pale. It gave a philosophical horizon to chemistry without concerning the chemist in his daily work in the least. Dumas adopted another solution, which rather recalls Jean Philippe de Limbourg's distinction between the two points of view, physical and chemical. He presented two atoms to his students: the physicists' atom, which answers the need for a deductive, mechanical science, and the chemists', which is enrolled in an experimental project to characterize every substance arithmetically (Dumas, 1837, lessons 6, 7).

If the idea of the atom sparked discussions and debates, that of the equivalent or atomic weight became an incontrovertible idea, indispensable to all chemists. First of all, it offered a numerical value with which to identify positively and precisely the various simple bodies and at the same time to establish comparisons among them. It also provided a way of translating laboratory analyses into formulas. And, finally, it gave manufacturers a relatively reliable test for controlling operations in factories. Within a few years, all chemists were concentrating their efforts on this idea. As if it had been planned that way, their research converged toward the same objective: to determine with acceptable precision the value of atomic or equivalent weight for each known element.

To the question "what is acceptable precision?" the answer seems to have been almost unanimous: Berzelius. His *Essay on the Theory of*

Definite Proportions and the Chemical Influence of Electricity, which appeared in Stockholm in 1818 and was immediately translated into several languages, was taken as the standard reference by most chemistry professors. His atomic weights became the authority to the extent that everything that preceeded them was wiped out. After Berzelius, many textbooks would not refer to Wenzel and Richter; they would hardly mention Dalton, but they would reuse all of Berzelius.

How was a short essay able to capture all the knowledge of an epoch? This phenomenon of immediate acceptance and appropriation, comparable to the one following the reform of nomenclature in 1787, had to do primarily with Berzelius's new symbolism. He replaced Dalton's notation of circles and points by letters. The initial or the first two letters of the Latin name of each simple body represented the equivalent weight of that substance, and Berzelius introduced exponents to indicate the number of times the atomic-weight value designated by the symbol was present in the compound. This notation is still in use today. The few modifications it has undergone consist of the suppression of certain graphical conventions devised by Berzelius to shorten the formulas[16] and the transformation of his exponents into subscripts.

Berzelius's authority also derived from the project he started in 1818 and continued throughout his career: to correct Dalton's atomic weights. He did a series of analyses, syntheses, precipitations, and other experiments on nearly 2,000 bodies—mostly oxides and salts—to determine their exact composition and the equivalent weights of their constituents. His "Table of Atomic Weights," published in 1818, was repeatedly revised and republished by Berzelius himself. He prided himself on never using a numerical value that he had not established or at least verified himself. Berzelius's strict standards had a double benefit: they became a guarantee of precision, and his table served as an international standard until about 1835–1840; they inspired other chemists, such as Jean-Baptiste Dumas, Jean Servais Stas (1813–1891), and Jean-Charles Galissard de Marignac of Switzerland (1817–1894), who tried to outdo him in precision, using every possible means.

An Arsenal of Laws

Chemists were not content to use the laws of proportions to determine atomic-weight values. They used every possible resource, every bit of knowledge and know-how, from the study of gases to crystallography, by way of the theory of heat and various measuring techniques—gravimetry, volumetry, calorimetry, geometry, and goniometry—and all the available instruments to determine the densities of gases. By mobilizing scientific potential around this well-defined objective, they were able to cover the largest possible territory, including electricity, heat, and crystallography in the 1830s.

The first law to fashion a network of experimental facts from different disciplines was formulated in 1811 by Amedeo Avogadro (1778–1856), a physicist from Turin, and then announced independently by André Marie Ampère in 1814. "To explain the fact discovered by Gay-Lussac," Avogadro suggested that "in the same conditions of temperature and pressure, equal volumes of different gases contain the same number of molecules." It was a simple hypothesis to explain the conjunction of gravimetric and volumetric proportions. But this hypothesis implied another. In effect, in the case of gaseous combination, there was a contradiction between two requisites: on the one hand, a molecule consisting of two or more elementary molecules had to have a mass equal to the sum of the molecules; on the other hand, the number of composed molecules had to remain the same as the number of molecules in the first body. The difficulty was explained by assuming that "the constituent molecules of any simple gas . . . are not formed from a single elementary molecule but are the result of uniting several molecules into a single one by attraction, and that when molecules of another substance must join with the former ones to form complex molecules, the integrating molecule that should be the result divides into two or more parts" (Avogadro, [1811] 1991, p. 96). For example, water would be formed from a half-molecule of oxygen and two half-molecules of hydrogen.

From our present perspective, this was a formidable step forward. With the distinction between atom and molecule, the first being the agent in the combination and the second the subject of the reaction, the essential basis of the atomic theory of chemistry was established. But this law, to us a brilliant idea, was ignored or rejected by the majority of chemists until the 1860s and even longer. How do we understand such obliviousness? Blindness? Inertia? Resistance to innovation? A scandal to the chemist, a vital question for the historian, this delay is explained by several considerations (Brooke, 1981; Fischer, 1982). In spite of the solution that it offered for Dalton's objection to Gay-Lussac, the chemists' reception of the hypothesis of di-atomic gases was cool. It seemed to them scandalous to imagine molecules consisting of two elementary molecules or atoms of the same nature. They could easily imagine molecular structures formed by the attractive union or affinity of two different atoms. But the union of two like atoms into one molecule appeared impossible, inconceivable, especially in the context of Berzelius's electrochemical theory, in which every combination was explained in terms of opposite electric charges.

Over and above the phenomena of theoretical resistance, this hypothesis was not really indispensable at the time it was formulated. To a chemist occupied with isolating and defining new bodies or perfecting this or that procedure, the distinction between "elementary" and "integrating" molecules of simple bodies introduced a superfluous complication. Chemists were even less inclined toward it because Avogadro's hypothesis was not very convenient in determining atomic weights, since only a minority of the known substances could be treated in the state of vapor or gas. Finally, Avogadro's law was only one law among others that also provided tools for determining atomic-weight values.

Dulong and Petit's law of specific heat was based on another field, calorimetry, which had already been explored by Lavoisier and Laplace. Pierre Louis Dulong (1785–1838) undertook his study of specific heat in order to confirm the atomic hypothesis in spite of the opposition of Berthollet (his teacher at the Polytechnique). Dulong and his young colleague, Alexis Petit (1791–1820), measured the specific heat (i.e., the quantity of heat needed to raise the temperature of one gram of a substance one degree Celsius) of thirteen simple bodies by the cooling method, using the specific heat of water for the unit. They then determined the heat capacity of each atom, by taking the product of the atomic weight and the specific heat, and obtained nearly the same values (between 0.3830 and 0.3675) for the thirteen bodies tested. From this they concluded in 1819: "the atoms of all simple bodies have the same heat capacity." This law could not lead directly to atomic-weight values, because it presupposed them. But it did provide a verification and the ability to choose one value among the several values possible for an element. In this way Dulong and Petit modified several of Berzelius's atomic weights before Victor Regnault showed that the law was only an approximation and in turn corrected some atomic-weight values in 1840.

Mitscherlich's law of isomorphism, published between 1819 and 1823, was determined by the use of crystallography to calculate atomic weights (Mauskopf, 1976). Professor Eilhard Mitscherlich of Berlin (1794–1863) studied bodies of different chemical composition that crystallized in identical or very similar forms—like natural lead sulfide and sea salt, for example—and called them isomorphs. These bodies could substitute for each other in a crystal without changing its form (except for some variations in angles) and crystallize together whatever the proportions. Mitscherlich immediately confronted the question of the link between this property and the atomic hypothesis, which he designated, following Berzelius's lead, the "theory of chemical proportions." He postulated that only the number of atoms determined the crystalline form, and he attributed the little angular variations to the nature of the atoms. Using the analogy of the chemical composition of isomorphic bodies, one could then, knowing the equivalent weight of the atoms that compose one, determine that of the atoms in the other. In this way, having demonstrated the isomorphism of sulfates and seleniates, Mitscherlich determined the atomic weight of selenium from that of sulfur.

Unlike Avogadro's law, Mitscherlich's did not require a hypothesis on the molecular composition of bodies, but its field of application was limited to crystallizable substances. Like Dulong's and Petit's law, Mitscherlich's presumes at least the knowledge of the atomic weight of one term in the isomorphic series, and therefore it could not be an exclusive method of determining atomic-weight values. So it was a network of laws formulated within ten years of Dalton's hypothesis that provided a group of complementary methods for determining the numerical values of atomic weights.

Doubts and Digressions

The drama began when complementarity turned to discord. Dumas, one of the rare chemists to use Avogadro's law, came across a glaring contradiction in 1832. The vapor densities of sulfur, phosphorus, arsenic, and mercury were two or three times greater than the values indicated by specific heat and the chemical analogies.[17] What to choose? Either maintain with Avogadro that equal volumes of gas or vapor contained the same number of atoms and adopt barbarous, from the strictly chemical point of view, formulas such as $H_2S_{1/3}$ for hydrogen sulfide, or $H_3P_{1/2}$, or else use the atomic-weight values determined according to analogies and specific heats (32 for sulfur, 200 for mercury, assuming that mercury vapor had two times fewer atoms than an equal volume of hydrogen, and sulfur vapor three times more).

To resolve this dilemma, Dumas declared that the proportionality of atomic weight to gas density seemed to be defective in cases where a gas was a simple body. He retreated to the law of isomorphism, which seemed more reliable, and condemned Avogadro's law, which fell into disrepute and then obscurity. By a sort of chain reaction, Avogadro's law pulled Gay-Lussac's law and the Daltonian atom into disgrace along with it. Thénard, who also rejected Avogadro, still left a chance for the atom: "Atomic chemistry would be a purely conjectural science if it had to limit itself to this sort of consideration." But in 1836, Dumas did away with the whole mess. He concluded one of his *Leçons de philosophie chimique* at the Collège de France: "What remains of the ambitious excursion we allowed ourselves into the domain of atoms? Nothing, at least nothing necessary. What remains is the conviction that chemistry lost its way, as usual when, abandoning experiment, it tried to find its way through the mists without a guide. Using experiment you will find Wenzel's equivalents and Mitscherlich's equivalents, but you will never find the atoms that your imagination dreams up . . . If I were master, I would erase the word 'atom' from the science, persuaded that it goes beyond experiment; and in chemistry we should never go beyond experiment" (Dumas, 1837, p. 249).

Was Dumas the real master? In 1844 equivalents had totally eclipsed atomic notation in the *Annales de chimie*. Most French chemists firmly resolved to restrict themselves to equivalents. Several influential chemists in Germany—Liebig, Wöhler—relied equally on a system of equivalents conceived by a German chemist, Leopold Gmelin (1788–1853), in a text circulated all over Europe. This system was constructed exclusively on the weight proportions of combination, without reference to volumetric proportions. Although some chemists had adopted Berzelius's formula, H_2O, Gmelin returned to HO, in the name of a wise principle: when in doubt, choose the simplest and most elegant formula. The retreat back to equivalents was often accompanied by a distrust of physical methods and a renouncement of the attempt to reach the reality beyond the phenomena in chemistry. And chemistry prided itself on sticking to the heroic humility of a positivist science.

Speculations

Chemists had certainly made a vow of chastity with regard to speculation about atoms, but that did not keep them from flirting with an even more speculative hypothesis formulated early in the nineteenth century by an English physicist, William Prout (1785–1850): that the multiplicity of simple bodies, which were becoming more numerous every day, resulted from a single, original element, hydrogen. This thesis got unexpected support when Dalton chose hydrogen for the unit in his system of atomic weights. One could thus hope to test the hypothesis experimentally by showing that the atomic-weight values of the other elements were whole-number multiples of that of hydrogen. This project was undertaken by Thomas Thomson, Dalton's and Prout's champion, in his Glasgow laboratory. But when he published his atomic weights in 1825, they were criticized by Berzelius, who reproached Thomson for eliminating the bothersome decimals. This resulted in a strong impetus to push the precision of atomic weights farther and farther to refute Prout's hypothesis. In 1831 Prout riposted: the atomic weights of all the elements must be whole multiples of a fraction of hydrogen. Set up this way, the hypothesis was hardly susceptible to falsification any more, but at least it no longer encouraged the rounding off of atomic weights. Then it became a great success on the continent, where it found illustrious defenders in the persons of Jean-Baptiste Dumas and Jean-Charles Galissard de Marignac.

How can we explain the attraction of such a daring hypothesis when chemists refused to talk about atoms? Why lend any significance to the arithmetic values of atomic weights when the unit $H = 1$ was purely conventional? Prout's success was a symptom of a paradox hidden in the analytical program. In a sense the atom had completed the chemistry of simple bodies in a harmonious way, atomic weight reinforcing the individuality of the chemical element by a quantified property, measurable with precision. But in other respects nothing prevented the application of Lavoisier's analytic program to the atom itself and the attempt to divide and subdivide it. Even better—everything encouraged this: doubt about the absolute simplicity of simple bodies and their indefinite number maintained a profound incertitude about the definition of a basic, apparently elementary, concept, the chemical element. The unanimous agreement on Lavoisier's definition left an ambiguity between element and simple body, which was exploited and reinforced by the analytic achievements of nineteenth-century chemists. They thus reactivated the search for the unique, original and ultimate element beyond the multiplicity of simple bodies.

Prout's hypothesis encouraged not only the race for atomic weights but also attempts at classification as a function of atomic weight. Establishing correspondences between the arithmetical ratios of different atomic-weight values and the chemical analogies of elements was like discovering familial relationships and indices of consanguinity. Classifying the elements was like constructing a genealogical tree for the nonliving material world.

So atomic weight was accepted as a criterion of classification from 1817 on. On the basis of Berzelius's atomic-weight values, Johann Wolfgang Döbereiner (1780–1849), professor at Jena, established a series of "triads" of elements that already incorporated the idea of a correlation between the arithmetic of atomic weights and the analogies of chemical properties. On the same basis, Leopold Gmelin outlined a more general system of elements a few years later (Spronsen, 1969).

Although most chemists focused their attention on atomic weights, there were profound disagreements about their meaning and significance. An American disciple of Prout, Carey Lea, imagined a system of chemical elements with negative atomic weights. Taking a different perspective, Marc-Antoine Gaudin (1804–1880), a convinced atomist, pointed out in 1831 the periodicity of certain properties—volatility, fusibility—on the basis of the relative number of atoms in a given volume. In the one case, the atom was a numerical value devoid of physical reality, no more than a code for deciphering genealogical relationships; in the other, the atom was a physical entity with spatial reality. Thus, far from resulting in a complete and coherent system, the analytical program outlined by Lavoisier and relaunched by the focus on atomic weight was supported by a variety of divergent theoretical interpretations.

Around 1840, analysis, having become a routine professional practice, would cease to be the focus of chemists' efforts. That did not signal the end of its period of fertility. It would be revived in the 1860s as a result of a new method, spectral analysis.[18] The principle of spectral analysis—each group of spectral rays is characteristic of an element—was immediately put to use by its inventors, Gustav Kirchhoff (1824–1887) and

Robert Wilhelm Bunsen (1811–1899), to identify two new alkaline metals, cesium and rubidium, which received their names from the colors of the rays in their spectra. This was followed by a new wave of simple bodies in the 1860s, which underlined the urgent need for a classification scheme. But in 1840 a new chemical theory came along to challenge Berzelius's dualism and at the same time open a new field of research, a theory concerning a new activity for the chemist—substitution.

Substitution: A Source of Conflict

The replacement of one element by another in a compound (for example, an atom of hydrogen by an atom of chlorine in a hydrocarbon) would turn the whole landscape of chemistry upside down in the 1840s. It was a small cause with large effects. A new research program originated from substitution—to recognize constants while multiplying variations—that would inspire a new discipline, organic chemistry.

At the beginning of the nineteenth century, in spite of the disappearance of the traditional rubrics "vegetable chemistry" and "animal chemistry," the textbooks continued to discuss organic compounds of vegetable or animal origin and sometimes the nutrition of plants and the respiration of animals. Nobody questioned their being part of analytical chemistry. Had not Lavoisier himself extended his work to organic compounds? Burning coal, alcohol, oils, and sugar with a known quantity of oxygen, he weighed and characterized the products formed. The fermentation of wine, which would become the focus of disputes between chemists and biologists, was so perfectly integrated into Lavoisier's chemistry that it was in the chapter on fermentation in his *Treatise* that he formulated the famous principle of conservation and set down the first reaction equations. Lavoisier's collaborator Antoine Fourcroy described in his *Système de chimie* (1800) a set of eight analytical techniques for the identification and determination of the quantity of organic substances. He then related the whole spectrum of vegetable and animal acids to four fundamental constituents: carbon, hydrogen, oxygen, and nitrogen.

By what mysterious process did the title "organic chemistry," which then meant the chemical study of organic bodies, receive its current meaning of the chemistry of carbon and its derivatives? Was this a second chemical revolution?

In a sense, substitution, which caused this semantic shift, appears to have been a kind of revolution, because it introduced new concepts, new theories, and new practices, as Lavoisier's chemical revolution had done. The burden of chemical explanation was displaced from the simple body and the element to groups of atoms and molecular architecture. But this event was not felt as a break, a "rupture" by the actors. Rather, it acted as a force for revising and reorganizing chemical knowledge. In the daily course of normal science, the whole discipline was silently redefined through a number of controversies without causing a crisis.

The Chemistry of Organic Beings

Substitution posed the question of the identity and unity of chemistry again. At the beginning of the nineteenth century, the unity of the discipline had been assured because the laws of inorganic chemistry extended to compounds produced by organized beings. In the 1850s the situation reversed itself. It was organic chemistry that provided concepts with which to study and classify the inorganic realm. In twenty or thirty years—hardly the length of a career—chemical theory was reorganized into a consistent new system, one that no longer proceeded from the simple to the complex.

In fact, this arrangement was ill-suited to the organic domain. It was manifestly impossible to explain the individual properties of sugar or acetic acid, say, by the nature and proportion of the four elements (carbon, nitrogen, oxygen, and hydrogen) common to all. The ancestral techniques of apothecaries or perfumers, who carefully extracted the essences of perfumes, turned out to be more relevant here than the style of analysis preached by Lavoisier. The elementary analysis that led to simple bodies could evaluate substances only after having destroyed them, while the analysis of perfumers, which stopped at "proximate principles," posed the question of their rich multiplicity and perhaps that of the internal arrangement of elements that could be responsible for it.

The interest in such analysis is illustrated by the work of Michel Eugène Chevreul (1786–1889) on animal fats. Chevreul showed that all animal fats consisted of variable quantities of three fatty acids—stearic, "margaric," and oleic—and he later isolated and characterized a number of fatty acids (Chevreul, 1823). These were extremely interesting products for industry, and Chevreul himself invented the stearin candle, which replaced traditional wax candles in the nineteenth century, after further work had made it profitable (Emptoz, 1991). To separate the proximate principles of plants and animals, very delicate techniques were required. Chevreul's method began with the preparation of a soap, which was in turn separated into a solid and a liquid. The liquid was then decomposed and distilled until it produced a "volatile acid" and glycerin (Chevreul, 1824). A tradition of fine analysis different from Lavoisier's tradition but just as important for industry developed from Fourcroy to Vauquelin to Chevreul.

How did chemistry texts define proximate principles or compound radicals? To Chevreul they were isolable substances with well-defined physical and chemical properties. To Berzelius radicals were united two by two in compounds, exactly as elements and groups of elements entered into the formation of mineral salts, and, like simple bodies, they could be isolated as residues of analysis. Organic radicals therefore bent to the logic of mineral chemistry. Besides, Berzelius extended the dualism and the categories of acid, base, and salt to organic compounds. On the model of ammonium oxide—written N_2H_8O—produced by the addition of one equivalent of ammonia and one of water, he defined ether as the oxide of a radical, C_4H_{10}, and wrote it as C_2H_5. He wrote the composition of acetic acid as $(C_4H_6)O_3 + H_2O$, on the model of sulfuric acid $(SO_3 + H_2O)$.

If the extension of the categories of inorganic chemistry to organic chemistry assured a certain coherence among them, it also blurred the demarcation line between them. The animal or vegetable origin of compounds seemed to weaken as a criterion during the 1830s, but no other standard suggested itself. Neither the number nor the nature of the constituents was adopted unanimously. In our historiographical view it is clear, however, that organic chemistry did not cover the chemistry of carbon in the early nineteenth century. For Berzelius, organic chemistry was still an integral part of physiology, whose aim was to describe the composition of living beings and the chemical processes involved therein. The "particular combinations formed by carbon, nitrogen, oxygen, and hydrogen" was a rather marginal chapter of his treatise, an appendix to inorganic chemistry. Dumas went so far as to envision eliminating the distinction between organic and mineral chemistry in 1834. And in 1835, in the lessons devoted to chemistry in his *Cours de philosophie positive*, Comte decried the inconsistency of organic chemistry; he divided its domain between a part that should rightfully belong to inorganic chemistry, since its subject was more complex compounds, and another that rightfully belonged to physiology, since it studied phenomena related to the living (Comte, 1835, vol. 1, pp. 637–649).

A Crystallographer's Approach

More than through the life sciences, the redefinition of organic chemistry came about through the mediation of a neighboring discipline, crystallography. Since the beginning of the century crystallographers had been inviting chemists to consider the shape and arrangement of atoms, to direct their attention to molecular structures. In characterizing each type of mineral by a fixed polyhedral form, René Just Haüy (1743–1822) postulated a correlation between the macroscopic properties of a substance and the microscopic ones (Metzger, 1918; Mauskopf, 1976). He stated that the form of the "integrating molecule" was determined by those of the "elementary molecules" that composed it. In sum, he considered the molecule as a structural unit subdivided into more elementary units. This was the hypothesis that the crystallographers gave to the chemists, along with a precious instrument—isomorphic crystals that revealed the identity of a crystal structure by the variation in its constituent elements.

Several French scholars, influenced by Haüy's teaching, tried to apply the crystallographic notions of molecular geometry to the study of chemical composition. In this way Ampère had already arrived at the same hypothesis as Avogadro on the division of gaseous molecules when they enter into combination. Marc-Antoine Gaudin, suggesting the idea of an "architecture of the world of atoms," hoped to find the key to chemical phenomena in the spatial arrangement of atoms inside molecules (Gaudin, 1833), and Alexandre Edouard Baudrimont attempted an interpretation of chemical combination in terms of the rearrangement of atoms (Baudrimont, 1833).

A doctoral student who studied crystallography at the School of Mines, Auguste Laurent (1807–1853), had the idea of opposing this crystallochemical approach to the electrochemical interpretations of combination.[19] In 1836 Laurent was preparing his doctorate in chemistry in Dumas's laboratory with a dissertation entitled *Recherches diverses de chimie organique* (Jacques, 1954). He concluded from several experiments that some hydrogen atoms were driven out and replaced by oxygen or halogen atoms.

This result was not revolutionary in itself, for it had been well estab-

lished by Dumas. Dumas once told a colleague that during a party at
the Tuileries in the 1830s the candles were releasing a particularly irritating smoke. Having been asked by his father-in-law, the chemist Alexandre Brongniart, to study this problem, Dumas showed that the
vapors were hydrochloric acid, which resulted from the replacement of
the hydrogen atoms in the wax by chlorine atoms during the bleaching
of the wax with chlorine. This anecdote provides an example of substitution, which Dumas compared to other cases already noted by Faraday in the action of chlorine on ethylene in 1821, by Gay-Lussac on
cyanogen in 1823, by Liebig and Wöhler on benzaldehyde in 1832, and
finally by Dumas himself on terebenthine essence, alcohol, and acetic
acid. Dumas interpreted all these reactions in 1834 with a "theory or
empirical law of substitutions," sometimes called "metalepsy" ("exchange" in Greek), which can be summarized as follows: when a substance containing hydrogen is submitted to dehydrogenation by
chlorine, bromine, iodine, oxygen, etc., for each hydrogen atom it loses,
it gains an atom of chlorine, bromine, iodine, or half an oxygen atom.
In Dumas's view it was simply one more empirical law, nothing to
threaten the theory of combination.

But the young Laurent, writing his dissertation, was not content simply to describe the phenomenon. He published a memoir in the *Annales
de chimie* that, even before he had reported any experiments, launched
a theoretical controversy (Laurent, 1836). Laurent openly attacked Berzelius's dualist electrochemical interpretation of combination. The substitution of an electronegative element, such as chlorine, for an
electropositive one, such as hydrogen, should change the nature of the
substance totally; this was not in fact the case, since the starting material
and the chlorinated product had similar properties. The dualistic view
of combination as the addition of two radicals that could be isolated by
analysis was equally irrelevant for interpreting such reactions.

For Laurent, who thought like a crystallographer, in terms of the
spatial arrangement of atoms and molecules,[20] a compound was not a
simple juxtaposition but a unitary structure built up by progressive
substitutions in a starting pattern called a nucleus. This model entailed
a displacement of chemistry's center of interest: the molecule, and no
longer the atom, became the significant unit in chemical reactions. This
would be confirmed by the definitions later proposed by Laurent in his
Méthode de chimie: the atom represents the smallest part of a simple

body that can exist in combination; the molecule represents the smallest part of a simple body that can be used in a chemical reaction.

The great Berzelius responded with disdain to the attacks of this obscure student. In the report he published every year on the progress in chemistry as a whole, which was widely read and translated into several languages, he summarized Laurent's publication in 1837 and concluded: "I think it is not worth mentioning such theories in my future reports." The next year he indeed wrote nothing about Laurent's theory of the "nucleus." According to Laurent, all organic compounds derived by progressive substitution from a basic nucleus, C_8H_{12} (for C = 6), visualized as a four-sided prism in the style of the crystallographers: eight corners were occupied by carbon and twelve atoms of hydrogen were at the centers of the twelve edges. Laurent derived the halides from this nucleus by substituting chlorine for hydrogen, and then aldehydes and acids by adding oxygen.

Although he was the author of the law of substitution, Dumas at first rejected his young assistant's geometrizing speculations illustrated with rigid shapes. He reproached him publicly in his lectures at the Collège de France in 1836. It was true that chlorine took the place of hydrogen, but that did not mean binary radicals should be replaced by more or less probable, vague formulas (Dumas, 1837, p. 299). Laurent's immediate reply was to claim the right to go beyond the facts by invoking the memory of Berthollet. Dumas then insinuated that Laurent was a little unstable, carried away, and he disowned him: "I have never said that the new body formed by substitution had the same radical, the same rational formula as the first. I have said exactly the contrary on a hundred occasions. Let him who wants to claim this opinion support it; it is no concern of mine" (Dumas, 1837, p. 699).

While renewing his attachment to dualism, Dumas was pursuing his experiments with substitution. In 1838 he obtained crystals of chlorinated acetic acid by having one liter of gaseous chlorine act on 0.9 milligram of acetic acid in a flask exposed to sunlight. After having analyzed and characterized the product, which had properties very similar to those of acetic acid, Dumas wrote the reaction: $C_4H_4O_2 + Cl_6 = C_4HCl_3O_2 + 3HCl$ (for C = 6). By thus admitting that chloracetic acid had a formula similar to that of acetic acid, Dumas in turn broke with electrochemical dualism. He rejected Laurent's image of the "nucleus," however, and advancing the idea of "types" from which compounds were formed, he proposed a classification for those types. Sent away to Bordeaux, Laurent moldered away without support and without a laboratory. He complained in vain that Dumas did not acknowledge his priority and vanished from the international scene, while Dumas commanded attention as the champion of the theory of substitution. Only after his death in 1853 was Laurent recognized, thanks to Jean-Baptiste Biot's publication of his *Méthode de chimie*.

A Battle of Giants

Meanwhile, the fate of the unitary (i.e., antidualist) theory was played out in a merciless struggle among three great masters of the chemistry of the time (Leprieur, 1977). There had already been a few skirmishes between Dumas and Liebig on the nature of the benzoyl radical identified by Liebig and Wöhler and on the composition of ethers. From Stockholm, Berzelius arbitrated the fight with monarchical dignity. But when Dumas published his study on chloracetic acid, Liebig and Berzelius presented a united front against him. Berzelius defended his theory of binary radicals, writing chloracetic acid as an addition of carbon chloride and oxalic acid, $C_2Cl_6 + (C_2H_3 + H_2O)$. Leaving aside the electrochemical problem, Berzelius managed to save dualism in this way in the last edition of his *Treatise*, but at the cost of redefining all the compounds produced by substitution. The incredibly complex formulas and bizarre radicals this desperate attempt produced did not convince his contemporaries. No other chemist adopted Berzelius's last formulas, and dualism died with him in 1848.

   Taking the offensive, Liebig denounced the arrogance of "Dumas and consorts" and vowed to censure their works in his journal. Moreover, he published a caricature of Dumas's article on chloracetic acid under the insulting pseudonym of S. C. H. Windler. In spite of the ferocity of the polemic, the quarrel between Liebig and Dumas seems to have been mostly a dispute over words, for Dumas's "type" strongly resembled Liebig's definition of the radical in his *Annalen der Chemie* in 1838. According to that definition, a radical had to have at least two out of the three following characteristics: it had to be the constant constituent of a series of compounds; it had to be replaceable in compounds by a simple body; it had to be able to combine with a simple body or its equivalents. After five years of struggle, Liebig quietly changed over to Dumas's view.

In 1845 he would go so far as to dedicate the French translation of his *Letters on Chemistry* to him.

In retrospect it seems clear that the substitution of chlorine for hydrogen objectively condemned electrochemical dualism, and this impression is not purely retroactive: it no doubt explains the ardor and imprudence with which the young Laurent confronted his elders. Didn't he have the "phenomena" on his side? But the agreement between Laurent's judgment of the importance of the phenomenon and ours today does not add up to a "revolution" of the sort Laurent wanted to accomplish. The educational background, the academic position, and the authority and prestige of the protagonists of such a controversy are an integral part of it. That the alliance between a peculiar and perhaps marginal phenomenon and a fledgling scholar could upset a respected interpretive theory and force chemistry into a more or less speculative adventure would have been a very significant event for the discipline. Dumas's more modest proposition, the death of Berzelius, and Liebig's discreet change in attitude made the end of electrochemical dualism a "non-event," the death of an already degenerating research program in the sense of Lakatos.

The Theory of Types

A new leader for the French school of chemistry appeared at the denouement of this quarrel. Charles Gerhardt (1816–1856), former student and French translator of Liebig and a member of Dumas's laboratory, found himself in an uncomfortable position during the battle. But in the 1850s he went beyond the conflict in writing a *Traité de chimie organique,* which marked the beginning of the discipline in the modern sense and raised a new perspective on the chemistry of substitution: what status should be accorded to chemical notation?

Until now substitution has been described only in terms of the controversy it provoked. From 1835 to 1845, the debate was focused on the nature of chemical combination: whether it was the addition of two elements or two radicals united by electric charges, or the substitution of atoms on the basis of a nucleus or type. Another basic question was decided by reference either to the technique of analysis or to that of substitution: what is a radical—a residue of proximate analysis or an invariant structural arrangement of molecules?

For chemistry professors these theoretical questions translated into concrete and everyday terms: how to name, notate, and classify compounds. Carbon was primarily responsible for the inflation in the number of compounds. By the middle of the nineteenth century seven to eight thousand compounds had been identified. Ethers and aldehydes had been added to the more and more numerous alcohols. To guide the student through this labyrinth, rational, coherent formulas and an ordering principle were needed.

To create a system for classifying compounds on the basis of the theory of substitution was Gerhardt's essential object in his treatise on organic chemistry. He adopted the notion of the chemical "type," defined in 1850 by Alexander William Williamson, a professor in London and former student of Liebig in Giessen and then of Comte in Paris. Williamson thought all compounds were derived by progressive substitution from a fundamental type: water. Using water as the type, Williamson proposed a new way of writing radicals:

Water type	Alcohol radical	Ether radical
$\left.\begin{matrix} H \\ H \end{matrix}\right\} O$	$\left.\begin{matrix} C_2H_5 \\ H \end{matrix}\right\} O$	$\left.\begin{matrix} C_2H_5 \\ C_2H_5 \end{matrix}\right\} O$

The formula for acids was derived from water by substituting radicals for hydrogen. For the so-called polybasic acids, which could form several salts from a single body, Williamson imagined that the type "water" was condensed several times. And he characterized each radical by its "basicity": the number of atoms for which it could substitute—a sort of substitution equivalent.

Instead of relating all the compounds to a single type, variously condensed, Charles Gerhardt preferred three types. He adopted the ammonia type defined in 1849 by Hofmann and Williamson's water type, and in 1853 he added the hydrogen or hydrochloric acid type.

Hydrogen type	Water type	Ammonia type
$\begin{matrix} H \\ H \end{matrix}$	$\left.\begin{matrix} H \\ \\ H \end{matrix}\right\} O$	$\left.\begin{matrix} H \\ H \\ H \end{matrix}\right\} N$

With these three types Gerhardt interpreted a great many reactions, notably those in which the radicals were exchanged by double decomposition. He divided the organic compounds into three classes and even foresaw unknown compounds by substituting radicals for hydrogen in each of the types. He resisted any realistic representation of the internal architecture of the compounds and refused to think of the radicals as isolable bodies that were real and permanent. The radical was simply "the relationship according to which certain elements or groups of elements substitute for or transport each other from one body to another in double decomposition" (Gerhardt, 1853–1856, vol. 4, pp. 568–569).

The idea of relationship brought back to life in the middle of the nineteenth century a notion that had hit the jackpot in the eighteenth century in a previous attempt at organization and classification. Beyond the triumph of the chemistry of simple bodies and isolable substances, Gerhardt stuck to the relationship among bodies to construct a classification, as Geoffroy had once done. But the idea had become completely abstract. It no longer designated the affinity of one substance for another as identified in thousands of empirical reactions. Gerhardt's radical or relationship was nothing but a taxonomic scheme that revealed analogies, homologies. It did not represent a stable grouping of atoms, but only a group of compounds, one stage of classification. Its formulas had no ontological significance, so Gerhardt called them "rational formulas," and he admitted that one body could be given several formulas. Gerhardt asserted that one could know nothing of the real composition of substances. This led to the paradox that although structural chemistry was beginning to be organized and constituted into a discipline, it professed agnosticism on the subject of molecular structure.

In spite of their abstraction, structures played a real role in Gerhardt's theories. Types or radicals acted in a block, which was carried from one body to another. Each of them could be characterized by a substitution value. They behaved like units or elements. What is more, by making hydrogen a type, Gerhardt invited consideration of simple bodies not just as residues of analysis or as concrete, isolable substances, but as "radicals," or substitution relationships. In brief, the notion of types reopened the question of the explanatory function traditionally reserved for simple bodies.

Discord

The lack of all theoretical and practical ambition, which contrasted with Laurent's attitude, did not prevent Gerhardt from taking theoretical positions and provoking a long debate (Wurtz, 1879, pp. 59–65). While studying a large number of carbon compounds, Gerhardt had noticed disparities between the formulas currently accepted on the basis of equivalent weights and the quantities obtained in reactions. The collective retreat to equivalents in the 1840s, motivated by the difficulties encountered in the use of vapor-density methods, had totally eliminated considerations of volume from the scene. The equivalents corresponded, depending on the bodies, sometimes to two volumes (hydrogen, nitrogen, chlorine) and sometimes to one (oxygen). Keeping account of the volumes, Gerhardt noticed certain inconsistencies. Accepting with Berzelius that water contained two atoms of hydrogen for every one of oxygen, and carbonic acid one of carbon for two of oxygen, Gerhardt noticed that the quantities obtained from the reactions always corresponded to double the formulas. A minimum of H_4O_2 and C_2O_4 was formed. Gerhardt deduced from this that the accepted formulas were too strong, which complicated the notation unnecessarily. He therefore proposed to double the atomic weight of carbon (12 instead of 6) and of oxygen (16 instead of 8), and to reduce by half the atomic weight of certain metals. This allowed him to simplify a great many formulas, notably those of acetic acid, alcohols, aldehydes and hydrocarbons, by dividing by two. These simplified formulas went against any dualist notation and, on the contrary, lent themselves to interpretation in terms of types. For example, one could write sulfuric acid, H_2SO_4, and acetic acid, $C_2H_4O_2$, in the following ways:

$$
\begin{array}{cc}
\textit{Sulfuric acid} & \textit{Acetic acid} \\[1em]
\left. \begin{array}{l} SO_2 \\[1em] H_2 \end{array} \right\} O_2 & \left. \begin{array}{l} C_2H_3O \\[1em] H \end{array} \right\} O
\end{array}
$$

To convince chemists of the necessity for his reform, Gerhardt announced in the introduction to his *Treatise* that he would use Gmelin's equivalents in order to prove their irrationality. Was demonstration by absurdity more decisive than Gerhardt's authority, earned by his discovery of anhydrates and chlorides of fatty acids? In any case, the simplicity of the type formulas attracted a number of chemists. After Laurent, the first to rally, came Gustav Chancel and Wurtz in France, Edward Frankland, Hofmann, Williamson, Benjamin Collins Brodie, William

Odling, and John Hall Gladstone in Great Britain, and Kekulé and Adolf von Baeyer in Germany.

At that point the question of atomic notation became a drama—agreement was impossible. Until that time, in spite of conflicts, chemists had always succeeded in forming a broad consensus: on Berzelius's system in the 1830s, on Gmelin's in the 1840s. But in the 1850s the chemists' community became a Tower of Babel. Not only was there open conflict between equivalentists and atomists, but within each camp serious divergences had arisen. The equivalentists had to choose among equivalents based only on the gravimetric ratios of combination, or on volumetric proportions, or on substitution equivalents, or they could even opt for a mixed system by always choosing the simplest formula. On the atomists' side agreement was not complete either. Although Gerhardt's notation had served as a rallying point, it was nevertheless quickly reviewed and corrected by some of its adherents. In 1858 Stanislao Cannizzaro proposed doubling the atomic weights of a large number of metals again, and Adolphe Wurtz also introduced some modifications. There was a cacophony of figures and formulas. The same formula could be used to designate several different substances: for example, HO meant water for some and hydrogen peroxide for others; C_2H_4 was methane for some and ethylene for others. Inversely, the same substance could be written in different ways depending on the system: the best-known example is acetic acid, for which Kekulé turned up nineteen different formulas (Kekulé, 1861, vol. 1, p. 58). Decidedly, what at first had appeared to be the smooth and triumphant advance of nineteenth-century chemistry turned out to be a rather chaotic road full of potholes. Seventy years after the reform of the nomenclature, chemists were faced with a jumble of formulas and with differences in notation that made communication difficult, indeed almost impossible. Then August Kekulé decided to end the chaos and took the initiative by inviting his colleagues to a congress.

Reorganizing Chemistry

At the beginning of September, 1860, one hundred forty chemists from all countries, answering their colleague's call, arrived in Karlsruhe to discuss atomic notation. This would be chemistry's first international congress (Nye, 1983).

Chemists in Congress

The problem was clearly defined in the circular drawn up by Kekulé and Wurtz: to put an end to the "deep differences on words and symbols that harm communication and discussion, which are essential springs of scientific progress." This program gave the Karlsruhe congress a double dimension. On the one hand, it materialized the existence of an international chemical community and defined the rules for its functioning—namely, communication and the necessity for consensus. On the other hand, it raised a fundamental theoretical issue, for the agreement on figures and formulas was subordinated to an understanding on the definitions of basic concepts: atom, molecule, equivalent (Bensaude-Vincent, 1990).

Could the nature of atoms and molecules be decided by a convention, by a popular vote? In fact, three days of discussion did not produce agreement on these questions. The only measure agreed upon was the adoption of Berzelius's barred symbols. Elsewhere differences were obvious. Dumas expressed his nostalgia for the time when Berzelius's notation was an infallible guide. Kekulé declared that the only imperative was to choose a notation, no matter which one, and to stick to it. He nevertheless expressed his preference for a notation based on chemical considerations alone. Physical properties—volumes and gaseous densities—should not have priority. But the Italian chemist Stanislao Cannizzaro (1826–1910) made an effort to rehabilitate Avogadro's law. In a lively speech, he pushed his colleagues to admit the distinction between atoms and molecules and to adopt Gerhardt's system of atomic weights with occasional corrections.

Although the congress ended without having reached a consensus, Cannizzaro's plea convinced a majority of the participants. Everywhere in Europe, chemists began to use Gerhardt's notation as revised by Cannizzaro in their publications and treatises and to define the molecule and the atom as Wurtz did in his *Dictionnaire de chimie:* the atom was "the smallest mass capable of existing in combination," and the molecule was "the smallest mass capable of existing in the free state." Apparently the idea of a molecule consisting of two atoms of the same kind was no longer as shocking as it had been around 1820; in any case it did not seem to be an obstacle to using Avogadro's law. How was this obstacle overcome? By what mysterious process did chemists become accustomed to the idea of a bond between atoms of the same element? The reason was another contemporary discovery, crucial for synthetic chemistry, which we will discuss later in this chapter.

But the Karlsruhe congress was also the starting point for another story, for among the chemists enthused by Cannizzaro's arguments were two young professors, Julius Lothar Meyer (1830–1895) of the University of Breslau in Germany and Dmitri Ivanovitch Mendeleev (1834–1907) of Saint Petersburg. In the years to come each would elaborate a periodic system of the elements on the basis of the atomic weights advocated by Cannizzaro.

Mendeleev often said that the Karlsruhe congress was the first step in the discovery of the periodic law (Mendeleev, 1879). How are we to understand that? Did the new atomic weights give Mendeleev and Meyer an intuition of periodicity? In fact, both chemists originally became interested in classifying the elements for pedagogical reasons, in order to organize all the known facts in a textbook. Meyer constructed a first classification system in a treatise published in 1862, *Die Modernen Theorien der Chemie*. As for Mendeleev, finding no acceptable text except Gerhardt's treatise, which was devoted to organic chemistry, he decided to write his own text on general chemistry for his students. It was while he was editing his *Principles of Chemistry* in March 1869 that Mendeleev discovered the famous periodic law, which allowed him to classify all the known elements and to predict others. His classification was neither the first nor the only one (Spronsen, 1969). Mendeleev profited greatly from the many previous attempts that had pointed out numerical relationships among the atomic weights of families of similar elements. At almost the same time, others—Meyer, John Newlands, Odling—conceived very similar periodic systems and even left places empty for the unknown elements. After some dispute over priority, especially with Lothar Meyer, Mendeleev's classification finally eclipsed the others. How can we explain this success?

Let us reconsider the situation. Toward the middle of the century, the development of organic chemistry rocked the theory, the basic notions, and the notation of chemistry. Such a blow called for a redefinition of the relationship between inorganic and organic chemistry. Dualist theory was formulated within the study of mineral salts before being extended to organic compounds. Would the unitary theory elaborated in the study of organic compounds also extend to inorganic chemistry?

Perhaps, but other possibilities suggested themselves. Two possible ways of coordinating the ideas that emerged from organic chemistry with those of inorganic chemistry will be illustrated here by the classification systems elaborated by Dumas and Mendeleev.

The system proposed by Dumas was to "import" the notion of the radical into inorganic chemistry, following Gerhardt's example of treating hydrogen as a type. Sulfur could be redefined as a sulfurile radical; nitrogen, as a nitrile; phosphorus, as a phosphorile, and so forth. This extension had many advantages. It unified chemistry by removing an artificial barrier between its inorganic and organic branches: why should the chemistry of carbon obey laws different from those that ruled the other elements? It allowed the problem of classification to be viewed in its greatest generality, by treating simple bodies and compounds globally. And, finally, analogies between the families of the simple bodies (radicals) of inorganic chemistry and the radicals of organic chemistry created a strong presumption in favor of the complexity of the bodies thought to be simple; this was a powerful argument in favor of Prout's hypothesis. Dumas concluded likewise in his 1858 memoir, *Sur les équivalents des corps simples,* in which he showed that the arithmetic progression observed in the series of ether radicals was the same as that observed between the equivalent weights of several families of simple bodies (lithium, oxygen, and magnesium), as well as for the ammonium series and the halogen family.

Mendeleev's Table

Mendeleev took another route, equally inspired by Gerhardt. The article announcing his discovery to the Russian Chemical Society began thus: "Just as before Laurent and Gerhardt the words *molecule, atom,* and *equivalent* were used interchangeably, so today the terms *simple body* and *element* are often confused. Each of them does, however, have a distinct meaning, which it is important to specify in order to avoid confusion in chemical philosophy. A simple body is something material, metal or metalloid, with physical and chemical properties. The idea of the molecule corresponds to the expression 'simple body.' The name 'element' should be reserved to characterize the material particles that form simple bodies and compounds and determine the way they behave physically and chemically. The word 'element' calls up the idea of the atom" (Mendeleev, [1871] 1879, p. 693). Mendeleev illustrated the difference between element and simple body with the example of the element carbon, which comes in the form of three simple bodies—graphite, diamond, and charcoal—and with nitrogen, which is inactive in the free state but very active in combination.

The first lesson Mendeleev seems to have retained from the Karlsruhe congress was to introduce or reintroduce the distinction between atom and molecule sketched out by Avogadro and reformulated by Gerhardt and Cannizzaro. At first glance his distinction between element and simple body appears a little trivial. It had, however, not been clearly formulated before; the writings of most chemists remained ambiguous and did not specify precisely what they were classifying. So the distinction elaborated by Mendeleev entailed a complete reorganization of the conceptual landscape of chemistry.

At the beginning of his *Principles of Chemistry*, Mendeleev gave chemistry a new project: to explore, on the one hand, the relationships between simple bodies and compounds and, on the other, the elements contained in them. He met this objective thanks to the periodic law, which precisely defined a function between elements and compound bodies: "The properties of simple bodies, the constitution of their compounds as well as the properties of the latter, are periodic functions of the atomic weights of the elements, because these properties are themselves the properties of the elements from which these bodies derive" (Mendeleev, 1905).

As a consequence, the opposition between "simple" and "compound," which had been the main feature of Lavoisier's system, was pushed into the background for the benefit of a new concept of opposition between the empirical or phenomenal logical reality of simple or compound bodies and the abstract, underlying chemical element. While the "simple substance" was the key concept of a chemistry based on analysis, the element became the key concept and explanatory principle in Mendeleev's chemistry based on substitution. Only the element could explain the properties of simple bodies as well as of combinations. It was the element rather than the simple body that was responsible for the conservation of matter in chemical reactions, the element that circulated and was exchanged.

The displacement of interest from the concrete simple body toward the abstract reality of the element seems to have been an essential condition for the construction of a general system of the elements. Mendeleev, careful to point out the differences between himself and his precursors and rivals, stressed that before him, "because the bare fact always occupied the first place," one could at best form and group families, but one could not submit the multiplicity of bodies to a *general* law, much less predict unknown elements. Although this statement is unfair to some of his colleagues, it emphasizes an original aspect of Mendeleev's approach: his desire to find a natural, general law that suffered no exception. His model, analyzed at length in his *Principles,* is the law he called Avogadro-Gerhardt.

To state such a general law, Mendeleev had to leave out the exceptions. Barely having worked out the periodic relationship, he allowed himself to contradict a number of experimental results: he reversed the places of iodine and tellurium, in spite of their atomic weights; he doubled uranium's atomic weight; he corrected that of indium; etc. Moreover, in order to respect the periodic function, Mendeleev did not merely leave vacant places—which Newlands, Odling, and Meyer had already done—for three unknown elements that he named eka-aluminum, eka-boron, and eka-silicon, he also deduced their properties from those of the four elements that were their neighbors in the table. The precision of these anticipations of experimental discoveries illustrates the new epistemological status of the element: no longer a singularity isolated at the end of an experiment, the element was now an individuality defined by its relationships, by its place in a network.

Although it was as abstract as Gerhardt's type, Mendeleev's concept of the chemical element, which was both a condition and a product of the periodic classification, had a very real and not just a theoretical existence. For Mendeleev, the individuality of the chemical elements was an objective characteristic of nature as fundamental as Newtonian gravitation. Just as he refused to get embroiled in debates on the reality of atoms—for to him the relationship between atom and molecule was the most important—he made himself champion of the individuality and the multiplicity of the elements and fiercely fought Prout's hypothesis, which he considered a regression into the fantasy world of alchemy.

Ironically, Mendeleev's periodic classification was received by a number of his contemporaries as a decisive argument in favor of the reduction of elements to one original element. Marcellin Berthelot criticized it for that reason in *Les Origines de l'alchimie,* and the partisans of a unique, primordial element greeted it with enthusiasm. For many years Mendeleev protested against this use of his discovery and continued to proclaim his faith in the individuality of the elements. When he learned, toward the end of his life, of the discovery of radioactivity and the electron, he attempted the impossible in order to save his concept of the element. In a short essay, he advanced an explanation of radioactivity in terms of vortices of ether around the heaviest atoms (Mendeleev, 1904). Mendeleev treated ether as an element and placed it above the column of rare gases. He hoped not only to save the individuality of the elements but also to unify electromagnetism, mechanics, and chemistry into a new concept of ether—a grandiose but unfortunate hypothesis.

This error, which Mendeleev finally retracted, is, however, instructive, because it gives the modern chemist an unfamiliar view of the periodic system. It shows that the periodic system belongs to nineteenth-century chemistry, whereas in today's courses and chemistry texts the periodic classification is presented as an expression of the electronic structure of atoms, and Mendeleev, when he is mentioned, is portrayed as a precursor of the electron theories formulated in the twentieth century. Far from being a prophet of future developments in atomic theory, Mendeleev was trying to reorganize the knowledge of his time, and he drew a lesson from the chemistry of substitution to build up a broad system.

So the unity of chemistry, briefly contested, then menaced by the theory of substitution, seemed to have been rebuilt on a new foundation. The analytic logic reigning in the early nineteenth century had given way to a taxonomic logic of tabulation. Whether organic or inorganic, chemistry obeyed the same laws, and the constituent elements were classified in the same table. The distinction between the two branches seemed justified less by the nature of things than by pedagogical considerations. Correlatively, substitution lost its subversive status: it was no more than one mode of combination among others, specific only to the extent that it showed the difference between what was exchanged and circulated during a chemical reaction and what remained in a free state, the simple body or compound. Moreover, among the chemists assembled at Karlsruhe, some, like Kekulé, had already begun facing another problem, that of the arrangement of atoms in molecules, which would set them off upon a new investigative pathway: synthesis.

Writing Syntheses

"Chemistry creates its own object. This creative faculty, like that of art itself, distinguishes it fundamentally from the natural and historical sciences" (Berthelot, 1876, p. 275). This famous formula revealed a new identity for chemistry. While eighteenth-century chemists did their best to subordinate action to knowledge, chemical arts to science, chemists in the late nineteenth century distinguished their knowledge as a "creative faculty like that of art." The expression can be understood in two ways, the practical and the theoretical. Pure chemistry engendered applications; it led to the production of artificial objects imitating or supplementing nature, as Berthelot understood it. As Bachelard understood it, it took an epistemological meaning: the object of chemistry was not found in nature but constructed by the mind. Structures, radicals, and types were first and foremost theoretical constructions, so artificial substances produced on the basis of these structures were "materialized concepts" (Bachelard, [1930] 1973, pp. 63–78). The abstract/concrete ambiguity of the word *object* speaks to the double nature of synthesis as a practical as well as a theoretical activity.

Even though chemists were able to accomplish rather complex syntheses in the first half of the nineteenth century (such as that of urea, or that of acetic acid by Hermann Kolbe), it was only in the 1860s that synthesis became the focus of systematic research in which theory and practice moved in step with each other. Once certain ideas about structure that were indispensable to the start of this great work had been acquired, synthesis became a way of making new substances, a source of industrial profit, and a tool for exploring molecular architecture; it was a remarkable instrument of investigation. It took the baton from analysis and ran like hell with it.

What exactly did synthesis mean to nineteenth-century chemists? The broad meaning of the construction or composition of a substance had been specified according to various criteria (Russell, 1987, p. 169). It is now customary to distinguish among syntheses according to the nature of the product—either a natural substance reproduced in the laboratory or an artificial one—and this parameter is combined with the nature of the starting ingredients: either artificial substances or products extracted from nature. These distinctions, used for norms and patents today, were less frequent in the chemical literature of the nineteenth century than definitions according to the process of synthesis. Most chemists defined their work in relationship to a distinction popularized by Marcellin Berthelot, that between total and partial syntheses (Brooke, 1971). In the first case, substances were made from elements, such as carbon or hydrogen; in the second, substances were made from other, simpler compounds. Partial or indirect synthesis was successfully practiced in the 1820s, but the objective and ambition of many nineteenth-century chemists was total or direct synthesis. The difference between the two was not merely one of degree: the identity of chemistry was at stake once more. If the theory of substitutions had threatened chemistry with an internal cleavage, synthesis raised the problem of boundaries. Did the chemist have the power to make life? In nineteenth-century debates this was a burning question, so freighted with philosophical issues that it gave rise to a beautiful legend.

The Synthesis of Urea

Wöhler's synthesis of urea in 1828 was celebrated as an event of unprecedented importance. The first laboratory synthesis of a substance made previously only by living organisms, it was said to demonstrate the nonexistence of the vital force. This interpretation is a retrospective reconstruction of events by Hofmann (Brooke, 1968).

In reality, the vital force was not killed off by the synthesis of urea. For someone like Berzelius, it was sufficient to point out that it was not a direct synthesis from elements but a partial one from a cyanate. The cyanate itself was not prepared from elements but by oxidation of a cyanide obtained from the horns and hoofs of animals. Before and after 1828, Berzelius firmly believed in the existence of a vital force "that probably acts in the formation of organic combinations and that has been little observed until now." If Berzelius based this belief on the impossibility of accomplishing a direct synthesis of organic compounds, not all chemists agreed with him.

Liebig often declared that the primary object of chemistry was the artificial production of organic compounds. Even if the direct synthesis of sugar, morphine, and the like was not yet possible, it seemed probable to him that it soon would be. That is why in 1840 Liebig formulated an ambitious project to extend chemistry into the territory of physiology.

But he in no way based this project on the negation of the vital force as it was supposedly illustrated by the synthesis of urea by his friend Wöhler. Liebig did not question the existence of a vital force. The chemist's intervention into physiology did not imply a reductionist position. "Never will chemistry be able to make an eye, a hair, a leaf" (Liebig, 1845, p. 26). Liebig, like his friend Wöhler and their contemporaries, the thousands of readers of his *Letters on Chemistry*, carefully distinguished between two questions usually confused in the legends concerning the synthesis of urea: the doubtful specificity of organic compounds as possible objects of artificial synthesis, and the very reliable specificity of living organisms that are able to produce syntheses more complex than those achieved by the chemist in his laboratory with all his tools.

The synthesis of urea as a crucial experiment overthrowing a metaphysical dogma is thus a myth elaborated in a vague, positivist attempt to exalt the power of chemical synthesis. Far from refuting the vital force, the experiment provided Claude Bernard with an additional argument in favor of a clear boundary between chemistry and physiology: the chemist can imitate the products of living things, but he cannot duplicate nature's synthetic processes (Bernard, 1865; Bud, 1992).

If the synthesis of urea was not the end of the metaphysical belief in a vital force, how are we to understand the historical significance that contemporaries gave to this experiment? How did Wöhler and his colleagues perceive the synthesis of urea? In the conclusion of his article on synthesis, Wöhler did not discuss the vital force but the problem that would be called "isomerism." Urea was obtained from the same components as potassium cyanate, but it did not have the properties of this salt. What made this experiment such an important, extraordinary event in the eyes of Berzelius and Liebig was that it stated openly the problem of the arrangement of atoms.

In 1832 Berzelius's annual report declared that it was time to question the previously accepted axiom that substances consisting of the same constituents in the same proportions must necessarily have the same chemical properties. Actually, potassium cyanate and urea were not the only compounds with the same constitutional formula but different chemical properties. The fulminic acid studied by Liebig, the cyanic acid studied by Wöhler and the tartaric and racemic acids as established by Berzelius himself were other examples. One could no longer be content, therefore, with characterizing a compound by the nature and proportions of its constituents. Empirical formulas proved insufficient. Berzelius established a parallel between those compounds and other cases in which substances with a different elementary composition crystallized into the same forms (i.e., Mitscherlich's isomorphism). Pointing out the symmetry between the two phenomena, Berzelius first gave the name "heteromorphs" to bodies that contained the same number of atoms of the same elements but had different chemical properties and crystalline forms. Finally, to choose a more appropriate term he coined the term "isomers" (from the Greek *iso-meres*, "equal parts"). This generic term was quickly divided into subspecies: Berzelius called those substances that had the same constituent atoms with an equal relative number of them but different absolute ones "polymers" and used the examples of ethylene (or olefinic gas), which he wrote as CH_2, and an oil called *Weinöl* (C_4H_8). He called "metamers" substances with the same constituent atoms in the same relative and absolute numbers, but in different arrangements. As examples of metamers he used tin sulfate and sulfite, respectively written as $SnO \cdot SO_3$ and $SnO_2 \cdot SO_2$. Although they did not strictly follow Berzelius's terminology, chemists were aware that they had to deal with a new class of substances and phenomena, which obliged them to go beyond "raw formulas" to take direct account of the arrangement of atoms in the "compound atom."

Optical Isomers

The emphasis on atomic arrangement became even more urgent when the motley family of isomers was enriched by a new species, optical isomers. Louis Pasteur (1822–1895), a young graduate of the Ecole Normale, chose the subject of his doctoral research after reading a note published by Mitscherlich in 1844 on tartrate and paratartrate.[21] These two salts had the same composition, the same nature and number of atoms, the same chemical properties, and the same angles when crystallized. They looked identical in all chemical tests and even in the crystallographers' goniometer, except in one test: the polarimeter. They differed only by a physical property: tartrate polarized light, but paratartrate had no optic activity.

In 1848 Pasteur solved the enigma thanks to the link he made between the studies of the physicist Jean-Baptiste Biot on the rotatory power of certain crystalline substances and the work of crystallographers on the hemihedrism of crystals. Since a hemihedric crystal like quartz (with half its facets oriented to the right and half to the left) had optical activity, the tartrate should also show hemihedrism. Pasteur verified that it did. Then he worked on paratartrate, which had no optical activity, and therefore should not be hemihedric. Pasteur found the contrary. But he also realized that crystallized paratartrate is in fact a mixture of microcrystals with inverse symmetry. He separated the "right and left facets" manually and found that the two groups of separated microcrystals were optically active. It was therefore their mixture that must neutralize optical activity (Pasteur, 1860). Faithful to the tradition going back to Haüy, Pasteur related the behavior differences to the shape of the constituent molecules. These two crystals should be formed with asymmetrical molecules that could not be superimposed on their mirror image and that would later be called "chiroids" (from the Greek for hand) or "enantiomorphs."[22]

For a first attempt this was a master stroke! When Pasteur published his dissertation, Jean-Baptiste Biot took him in his arms, declaring "this discovery makes my heart beat faster." In fact, Pasteur was lucky, because examples of enantiomorphs that divide spontaneously in normal laboratory conditions and that can be separated manually are extremely rare (Jacques, 1992). In any case, this type of isomer suggested a possible link between the physical and chemical properties of a substance and the spatial configuration of atoms in its molecules. Pasteur was therefore forced to admit a distinction between atoms and molecules when he presented his results before the Chemical Society of Paris in 1860. But he did not mention it explicity and did not go to Karlsruhe in September to discuss the respective definitions of these terms with his colleagues. As we shall see, Pasteur was more interested in using molecular asymmetry as a clear boundary between the inert and the living than in the investigation of molecular structure. Others working in the field of research that he abandoned would invent stereochemistry, which would extend the possibilities of synthesis. But before discussing this new field, let us see how the synthesis program became possible.

From the Type to Atomicity

This program could only have been conceived on the basis of the "structural formulas," as distinguished from the "empirical formulas" used in analysis. The growing number of isomers identified indeed required that raw formulas be abandoned. The new notation indicating the arrangement of atoms in the molecule was based on an idea worked out in the 1850s: valence or atomicity (Russell, 1987).

In 1847, Edward Frankland (1825–1899), an expert in the analytical techniques that he had learned at Giessen with Liebig and used daily at the London Geological Survey, launched himself into the project of isolating the ethyl radical while working under Bunsen in Marburg. First he attempted to isolate it from ethyl cyanide with potassium, but instead of the desired radical he obtained bizarre and complex things. Next he tried to extract it from ethyl iodide, again with potassium, and he obtained a violent reaction with various hydrocarbons. Then he tried a less reactive metal, zinc, and he got an explosion with an abundant release of gas, in which he found zinc–ethyl iodide and diethyl zinc. So he had just discovered a new class of substances, the organometallic compounds, which was a concrete argument in favor of a reunion of organic and inorganic chemistry at the moment when the divorce was the most bitter. This was also the starting point for the theory of valence: it is as if, wrote Frankland, the zinc or tin or antimony atoms had just enough space to attract a fixed and definite number of atoms from other elements. He coined the term *valence* in 1852.

A year later, without any reference to Frankland, Kekulé formulated the problem in general terms, endeavoring to explain why the different elements combined in certain proportions rather than others. With atoms defined only by their atomic weight, Dalton could not account for that. Another intrinsic, individual property, the capacity for combination, should be given to atoms. That is why Kekulé substituted the term *atomicity* for valence.

Adolphe Wurtz (1817–1884), a professor in Paris, further elaborated and spread this notion. For him, it applied to radicals as well as elements: Gerhardt's hydrogen type was monatomic, the water type diatomic, and the ammonia type triatomic. Wurtz's definition of atomicity included both the capacity for substitution, which Williamson called "basicity," and the capacity for combination, called "valence." Ethylene, for example could substitute for two atoms of hydrogen in two molecules of hydrochloric acid and combine together the two remaining chlorine atoms. This capacity for combination and substitution could be used to define a substance and to predict its behavior according to the number of free and saturated valences.

Atomicity was therefore an instrument for prediction as well as for planning research, as illustrated by Wurtz's discovery of glycols in 1856. Having noticed a gap between the monatomic radical of alcohols and the triatomic radical of glycerine, Wurtz wondered if there were a diatomic radical. He identified it and showed that by saponifying it with silver oxide, one obtained a series of substances intermediate between alcohol and glycerine, which he called "glycols."

The structural formula expressed atomicity by indicating the number of valences exchanged among the different constituent atoms. Water, for example, was written $H^2 = O$. While Gerhardt's formulas could be used to classify bodies as a function of a basic structure arbitrarily chosen as a type or model, the structural formula showed all the possibilities of exchange or combination. Rather than expressing a hypothetical mode of formation of the compound, it indicated the various ways in which a synthesis could be made.

The road from the type to atomicity also corresponded to a displacement of interest from radicals toward elements. Although Gerhardt constructed his entire system on groupings of typical atoms and even conceived elements on the model of types, Kekulé gave the signal for a return to the elements. The project he fixed upon was to explain the properties of compounds not by radicals but by the nature of elements. Carbon was his main actor. The discovery of its tetravalency in 1858 influenced the entire chemistry of synthesis.

As a student, August Kekulé had reviewed much of the information and techniques available. At Giessen with Liebig, he mastered analytical methods; next he went to Paris to work with Gerhardt and became acquainted with the theory of types, which he came to understand more deeply during the time he spent with Williamson in London. Finally, he became a *Privatdozent* in Heidelberg with Robert Wilhelm Bunsen, master of the art of experiment and of the design of apparatus.

Considering the series of derivates of methane, then called "marsh gas," Kekulé postulated the quadrivalency of carbon, or the equivalency of four atoms of hydrogen. Then, while considering ethane and its homologues, he noticed that the number of hydrogen atoms never exceeded the limit indicated in the general formula C_nH_{2n+2} and postulated that carbon atoms could exchange a valence among themselves (i.e., combine with each other).

The affinity of a body for itself was a strange notion that seemed barbarous in Berzelius's time, as was shown by the hostility to Avogadro's hypothesis. The idea was extremely interesting, however, because it explained a variety of phenomena at once: the limit on the number of carbon atoms in saturated hydrocarbons; the fact that hydrogen atoms were always found in even numbers in hydrocarbons; the stability and the infinite variety of combinations of carbon. Here, finally, was the raison d'être of organic chemistry! The separation and the relative independence of carbon chemistry, which the theory of substitution had imposed as a matter of fact, had found a theoretical justification. A secondary benefit of the affinity of carbon for itself was that Avogadro's hypothesis of diatomic, gaseous molecules seemed less shocking. Now it is clear why the brilliant Cannizzaro had been able to convince a majority of the audience at the Karlsruhe congress in 1860. With the death of electrochemical dualism and the idea of the affinity of an element for itself, the principal obstacles to the acceptance of Avogadro's hypothesis had fallen away. From the 1860s on it would be referred to as Avogadro's *law*.

At the same time as Kekulé, a young Scottish chemist, Archibald Scott Couper (1831–1892), working in Wurtz's laboratory at the Faculty of Medicine in Paris, also postulated the quadrivalency of carbon. And Alexander Butlerov (1826–1886), a Russian chemist also staying in Paris in 1858, advanced a similar hypothesis. This simultaneity, like other historical cases of simultaneous, independent discoveries, raises the broad issue of determinism in scientific innovation. In the case of the tetravalency of carbon, it is clear that the specific properties suddenly attributed to carbon resulted from all the ideas advanced in the 1850s. But was it really the same idea that was put forward in 1858? Kekulé and Couper presented the same hypothesis on the basis of two very different philosophies. For Kekulé, a loyal disciple of Gerhardt, the symbols of the elements did not represent the atoms but only the magnitude of their valence. His rational formulas—drawn like strings of sausages—expressed reactions, not the actual constitution of molecules. Kekulé used the concept of atomicity without believing in the actual existence of atoms. On the other hand, Couper criticized Gerhardt sharply, saying that it was time to stop fooling around with ideal formulas and claiming to describe the way in which chemical combinations take place in a realistic way. Couper's boss, Wurtz, no doubt preferring Kekulé's attitude, did not back up his student. Couper's article in the *Comptes rendus de l'Académie des sciences* did not appear until later, and his career was compromised.

This dispute over priority shows a divergence in the meanings given to the atomicity of carbon. The differences became even deeper when it was necessary to explain why some elements do not always have the same capacity for combination, the same valence. Carbon itself seemed no longer tetravalent in carbon monoxide, CO, or in ethylene, C_2H_4. Couper related the variability in valency to two types of affinity in atoms: affinity of degree and elective affinity. Frankland proposed a distinction between latent and active atomicity. Wurtz developed a relativistic conception of atomicity: like affinity, it was a relative property of atoms—in other words, the saturation capacity of atoms was a reciprocal action, a sort of accommodation among atoms. Kekulé, on the contrary, regarded this variability as a scandal. Since valency was a fundamental property of elements, it should be constant, as atomic weight is constant (Kekulé, 1864; Hafner, 1979). Otherwise, this notion would be confused with that of equivalency. Mendeleev went so far as to proclaim the "failure of atomicity" and to declare that his periodic law would be preferable for constructing a theory of chemical combination. Finally, for Berthelot, equivalency and atomicity both had their advantages and disadvantages. So the disagreement over notation, submerged briefly by the Karlsruhe congress, resurfaced over the variation in atomicities.

The "miracle" of Mendeleev's table did not happen again with atomicity. Contrary to Kekulé's hopes, valency remained scandalously variable, and it would be necessary to wait for the quantum interpretation of the classification of elements for its variability to be linked to the electron configuration of atoms. But, as disappointing as atomicity was from the theoretical or philosophical point of view, it gave chemists a language that would open a new technique to them: identifying the arrangement of atoms in a molecule and constructing molecules with specific arrangements.

Building Molecules

For Marcellin Berthelot, the opponent of atomicity, synthesis was the inverse operation of analysis: a recomposition from elements. Transformations between the simple and the compound summarized all of chemistry (Berthelot, 1860). After Lavoisier, the "father" of analytical chemistry, came Berthelot, the champion of synthesis.

Berthelot presented an ambitious program of progressive syntheses approximately in these terms: first, combine carbon and hydrogen to form hydrogen carbonates, "which are, so to speak, the keystone of the scientific edifice." Next, synthesize alcohols, ternary substances formed from carbon, hydrogen, and oxygen. These in turn are the starting point for the third level: combine alcohols and acids to form ethers, or alcohols and ammonia to form "artificial alkalis" (strychnine, morphine, quinine, nicotine, and dyes derived from coal tar). Finally, the second half of the third level: oxidize alcohols. When oxidized carefully, they produce the aldehydes that bring out the most varied odors (mint, cinnamon, cumin, clove). Oxidize carelessly and you get organic acids, which are in turn the starting point for new syntheses: combined with alcohols, they produce esters; combined with ammonia, they produce amides. And here is urea, the frontier of living things. Q.E.D. Using only elements, one could reconstruct the whole chain of bodies. Chemistry could create everything, without flaws, without surprises. "It is enough to"—Berthelot repeated this phrase at each step of the grandiose edifice he was building, page after page, in a series of works inviting the reader to dream of the immense powers of science (Berthelot, 1860, 1876).

Nevertheless, in fifty years Berthelot accomplished only an insignificant part of this wonderful program (Jacques, 1987). He synthesized wine alcohol in 1854 from ethylene—not from elements; he obtained formic acid by combining carbon and soda. He recombined a fatty acid with glycerin to prepare the oils that he had previously analyzed. In effect he had set only the first stone in the projected building: the synthesis of hydrocarbons. In 1863 he made acetylene by directly combining carbon and hydrogen in his "electric egg." In 1867 he turned his attention to the acetylene polymers, which he wrote as C_2H_2. His object was to synthesize benzene, because it was "the keystone of the aromatic edifice."[23]

Benzene or Triacetylene?

Benzene, so named by Mitscherlich, who prepared and characterized it in 1831, was a case in which carbon's tetravalency seemed to be refuted. This only confirmed Berthelot in his determination to ignore structural formulas and to continue using raw formulas and equivalents. In benzene the weight ratio between carbon and hydrogen was the same as in acetylene. In other words, one liter of benzene contained the same elements as three liters of acetylene. The difference between acetylene and benzene must result from different condensations, Berthelot conjectured. This proportion, 12:1, was characteristic of all the aromatic compounds. He therefore heated acetylene in a bell jar to 550–600°C and, after several manipulations, obtained a yellowish liquid containing several polymers, which he separated by fractional distillation: 50 percent benzene, styrolene, etc. Berthelot summarized the whole thing in two formulas:

$$2C + 2H = C_2H_2$$

$$3C_2H_2 = C_6H_6$$

According to Berthelot's synthesis, benzene was triacetylene. The only acceptable formula was the expression of the empirical method of generation of a substance and not of a hypothetical structure, i.e., C_2H_2.

For a supporter of atomicity like Kekulé, the formula of benzene was quite different. Kekulé searched for years for a structure of this compound such that carbon's four valences would be satisfied. He groped around before arriving at the shape of a hexagon with alternate single and double bonds. In early 1865 he submitted a first conception of the structure of aromatic compounds to the French Society of Chemistry: a "nucleus" of six carbon atoms forming a closed chain, with lateral chains available for the formation of derivatives. The drawing showed the single and double bonds and marked unsaturated valences with a dot.

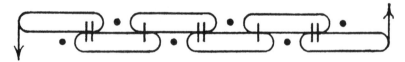

Kekulé's benzene formula

In a second memoir Kekulé showed a hexagon without locating the bonds. Finally, in a third article, published in 1867, he proposed a spatial model in the shape of a hexagon with alternating single and double bonds. From Kekulé's model one could predict a multitude of derivatives. The synthesis of aromatic compounds would no longer be a con-

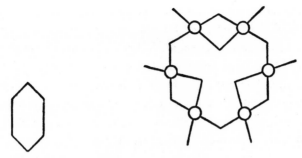

Kekulé's benzene models

densation of acetylene, like Berthelot's, but a substitution of elements or radicals for hydrogen atoms.[24]

Was this structure, which proved so important for the future of chemistry, constructed or dreamed up? Was it patiently elaborated, or visualized in a flash of intuition? Speaking at a ceremony in his honor in 1890, Kekulé said he owed his career to two visions: one revealed to him, as he rode a London bus in 1854, the carbon-carbon bond as a dance of atoms; the other, which came to him in Ghent in 1861–1862, as he sat before a fireplace, revealed the ring structure of benzene as a snake chewing on its own tail (Wotiz and Rudotsy, 1987). This so-called vision not only hid the laborious phase of exploration of the structure of benzene, it also eliminated the possible influence of the hexagon envisioned by Laurent and well known to Kekulé, who was thinking of translating the *Méthode de chimie*. Moreover, the legend of the snake elegantly sidestepped any claim by importunate rivals such as Joseph Loschmidt (1821–1893) or Archibald Couper, who had also come close to the structure of benzene. And in a period of extreme nationalist tensions, Kekulé's story was sure to please, because it placed the cradle of the entire chemistry of synthesis in a German brain, and that of a dreamer as well!

"Chemistry in Space"

In any case, Kekulé's students would know how to use their teacher's visions. Forgetting his reservations about the physical meaning of formulas and molecular architecture, they explained benzene's isomers, which Kekulé had been content to enumerate, from a realistic conception of the arrangement of atoms in space. Wilhelm Körner (1839–1925), Kekulé's assistant in Ghent, established the equivalence of the six hydrogen atoms and developed a method for determining the relative positions of the radicals in products of substitution. So in 1874 he was able to make the distinction between "ortho" isomers (in which two adjacent angles of the benzene hexagon are occupied by radicals substituted for hydrogen), "meta" isomers (in which two angles occupied by radicals are separated by a hydrogen), and "para" isomers (in which two angles occupied by radicals are opposite angles of the hexagon). Having distinguished the three isomers of dibromobenzene in this way, Körner proceeded to prove his hypothesis and identify the isomers. He substituted one of the four remaining hydrogen atoms by an NO_2 radical for each isomer. From the first of the three isomers he obtained three distinct, new isomers and concluded that he was dealing with the "meta" form; from the second, he obtained two isomers (i.e., the "ortho" form); and from the third, which could only be the "para" form, he obtained only one. Thus both Kekulé's hypothesis on the hexagon and Körner's own on the positional isomers of benzene were confirmed. Synthesis could therefore be an instrument for analysis in itself, since it provided a new means of identification and distinction. But it was also a project in itself: Körner managed to prepare 126 new derivatives of benzene. The chemistry of aromatic derivatives could thus become an area of systematic, planned research.

In 1874 another pupil of Kekulé, Jacobus Henricus van't Hoff (1852–1922), of the Netherlands, made the jump into three-dimensional space with his hypothesis that the four valences of carbon were located at the summit of a regular tetrahedron with carbon in the center (van't Hoff, 1887). The same hypothesis was announced the same year by a young French chemist, Achille Le Bel (1847–1930). Although they are always associated in present-day chemistry texts, Le Bel and van't Hoff arrived at the same result through very different creative processes (Compain, 1992).

Le Bel's starting point came from the lineage of Biot and Pasteur. He was seeking a general law to predict whether a substance in solution would or would not have rotatory power, would or would not turn the plane of light polarization. But, unlike Pasteur, who did not want to make a hypothesis on the shape of molecules, Le Bel leaned toward molecular geometry. He considered a body, MA_4, where M refers to a simple or complex radical (not necessarily carbon), and M is combined with four substitutable atoms, A. For derivatives from three, two, or one substituents, he predicted how many disymmetrical molecules should theoretically be obtained. Since, in the case of carbon, one obtains only a single optic isomer with two or even three substitutions, he concluded that the four atoms, A, occupied the summits of a regular tetrahedron (Le Bel, 1874).

Van't Hoff, as a good disciple of Kekulé, centered his entire investigation on carbon and called his theory "asymmetrical carbon." He started with the atomic theory and showed that it was not sufficient to account for certain cases of isomerism. He shaped the theory of the tetrahedron for predicting unknown isomers. In the context of the growing concern with carbon chemistry after the discovery of the structure of benzene, van't Hoff's point of view won out. In less than ten years, the theory of asymmetrical carbon was incorporated into German chemistry texts, in spite of the opposition of Hermann Kolbe, professor at Marburg, who denigrated structural chemistry and treated van't Hoff as a "Pegasus" flying off into the shameful space of speculation.[25] In France, on the other hand, the acceptance and spread of this theory—which obviously encountered opposition from Berthelot—would take longer. The result was that in Germany the reality of the atom would be fully admitted as a postulate, while in France it was contested until the beginning of the twentieth century and most students still learned chemistry using the system of equivalents.

The Art of Synthesis

While Berthelot was writing popular books glorifying the power of science, other chemists, starting with the benzene hexagon, devoted themselves to patient, tedious attempts at synthesizing new bodies through trial and error. In the sunny world of synthesis where Berthelot was the hero, "it was enough" to combine methodically and progressively. But Berthelot had never gone far in developing industrial processes. In fact, to obtain hydrocarbons homologous to benzene, it was *not* enough to heat and distill in Berthelot's way. In 1877 the Alsacian chemist Charles Friedel (1832–1899) and James Mason Crafts, a professor at the Massachusetts Institute of Technology, performed a reaction of a chloride or an alcoholic bromide with benzene by using a catalyst—anhydric chloride of aluminum. The alcohol radical substituted for one hydrogen of benzene, and the hydrochloric or hydrobromic acid was eliminated. For example, methyl chloride yielded methylbenzene or toluene. The synthesis of the whole aromatic series finally opened up, thanks to a new practice, that of the skilled chemist who knew exactly which catalyst would allow him to perform a specific transformation starting from two specified types of molecules.

In fact, to achieve the plan of synthesizing benzene derivatives according to the theoretical predictions, much more than simple structural formulas was needed. Once the project or the idea of a molecule was fixed, a perilous adventure began, as described by Roald Hoffmann, Nobel Prize winner in chemistry in 1981. The making of a molecule, Hoffmann argues, requires a strategy, which he compares to a chess game with nature (Hoffmann, 1991). One main constraint of the chemist's game is that he cannot act directly on materials. He has to delegate operations, to entrust them to intermediary molecules put to work in a flask. Their activity must be directed to a definite site in the molecular structure, to break a bond here, form another there. This requires constant tweaking and great skill, for each time a reagent is introduced, it tends to operate everywhere indiscriminately. If, for example, a reagent is used to add an atom of bromine or chlorine or break a double bond by oxidation, it does so wherever this is possible. The synthetic chemist therefore has to invent a device to limit its activity, design a pathway, and drive the reaction as a function of the available reagents. The precise order in which the different reagents enter the process has to be determined, the stages managed, and intermediaries with protective groups ("scaffolding" constructed to maintain certain pieces of the structure intact while the others are being worked on) created. It is an art in which delegation—letting reagents react—and manipulation—getting them to act where and as one wants—are required.

Moreover, to control what happens in the synthesis one has to vary the reaction conditions: change the temperature, the pH, the speed of the mixing process . . . Here again one must proceed step by step, without forgetting to test the intermediate products at each step to check their composition: dissolve, crystallize, make spectra—in brief, mobilize the complete artillery of analysis to control each step of the synthetic process. These strategies are simulated by computer nowadays, but that does not change the nature of the game fundamentally. The chemist advances his pawns on the chess board and attempts to obtain the predicted result: check and mate. To these already numerous and difficult tasks, industrial synthesis adds yet more constraints relative to

the security of manipulations and products, the profits and manufacturing costs of the process, and the patents held by competitors in the
marketplace.

We can now fully appreciate the difference between synthetic and
analytic chemistry. In Lavoisier's chemistry, control of a reaction concerned only inputs and outputs and consisted of a balance sheet. In
synthesis, the control of products remains important, but it is also
necessary to control the processes. The chemist cannot content himself
with the role of accountant; he has to act as a strategist leading troops of
various reactants into the battlefield. As the physicist or mathematician
needs a table of all integrated equations in his head in order to integrate
a new equation, the chemist needs to know all the processes already
mastered and all the available reactants in order to work out an elegant
new pathway. Each reactant, and often each catalyst that can transform
a molecule into a reagent, defines one possible route and, consequently,
an intermediate target as well. The conception of an entire reaction
chain may thus change with the discovery of a reactant or a catalyst. That
is why the development of a specific new reactant was an event important enough to be worth a Nobel Prize for Victor Grignard in 1912, for
what is today called a Grignard reaction or, more simply, "a Grignard":
"you do a Grignard and then . . ."

Organic chemistry texts usually present the classic, conventional reaction chains. But to the student and the researcher falls the problem of
directing the actors in a play, so to speak, and creating the situations they
need to achieve the desired goal. They must create a more or less original, more or less elegant "scenario" out of a whole set of possible routes
of transformation. Each great synthesizer develops his particular style, a
point that Hoffmann emphasizes by describing the making of molecules
as a creative process that requires all the faculties—reason, intuition,
esthetic taste. It is a difficult craft for mature, skillful, expert, dedicated
chemists—in sum, an art, a culture, a passion.

4

INDUSTRIAL EXPANSION

Heavy Chemicals: From Leblanc to Solvay

Seen from a long distance away, the air appeared to be tinged with reddish vapors and dust, which became acrid, irritating, nauseating as one approached. The houses in the area, more numerous every year, were huddled together with their backs turned to the soda works. But how could they ignore the enormous industrial complex deep in the valley, which sucked up a stream of sickly workers every morning? This clichéd description has become so banal that we forget how shocking were the upheavals brought on by chemistry at the beginning of the nineteenth century.

Let us listen to the story of a traveler in Provence around 1820: "Soon we arrived at the Septèmes Gorges, where, in one of the driest areas in dry Provence, several factories for making oxide and artificial soda had been constructed. The smoke that comes out of these laboratories blackens and burns the whole area; one would think oneself to be on the rim of a volcano. I asked a merchant about the results of this remarkable discovery. 'That's a good question,' cried a clergyman. 'Burn and destroy, that is the object and the end of all our innovations.' And starting to fulminate a burlesque curse on any improvement that did not go back at least half a century, he ranted against artificial soda, vaccination, and especially any mutual education system . . . Pandora's box, from which all the plagues escape, has been opened, and if this were not enough, from that seat of corruption they are threatening to establish another under the name of a chair in chemistry; but religious men are on to them; they have repulsed this insidious proposition and refused to give the money asked for such a perverse use" (de Jouy, 1822, pp. 159–160).

Clearly chemistry was damned. The artificial soda poisoned Provence all the more because it was associated with the events of Year II (during the French Revolution and until 1805, the French calendar followed a "revolutionary" system of numbering years). At the same time, it was perceived as a residue of the Terror, a vicious consequence of the Revolution.

Stories about Founders

Like many groups, chemists recount myths about their origins. Lavoisier is often celebrated as the founder of modern chemistry. The chemical industry also had a founder, a contemporary of Lavoisier, Nicolas Leblanc (1742–1806), the inventor of a process for manufacturing soda from sea salt. The two founding heroes shared a tragic destiny marked by the revolutionary upheaval: Lavoisier died on the guillotine in 1794, and Leblanc committed suicide in 1806.

What is the historical importance of the Leblanc process? In the eighteenth century, the soda needed for making soap, glass, paper, and dye was produced by burning vegetable matter—such as kelp, seaweed, and especially Spanish *barilla* (ashes of the soapwort plant). The cinders contained 20–33 percent carbonate of sodium. France was largely dependent on foreign sources, since it imported two-thirds of the raw materials for its production of alkalis. The idea of producing soda from a substance as common as sea salt was advanced as early as 1737 by Henri Louis Duhamel du Monceau (1700–1782), and several attempts at manufacture were made in Great Britain and France in the 1770s. In 1781 the Royal Academy of Sciences launched a competition that Nicolas Leblanc finally won in 1789, just at the beginning of the convulsions of the French Revolution.[1] On September 25, 1791, Leblanc obtained one of the first patents approved under the legislation on industrial property voted by the Assembly. It described how to make sea salt react with sulfuric acid in a large lead receptacle, whose cover was fitted with a pipe for releasing hydrochloric acid; and, then, how to mix the sodium sulfate thus obtained with charcoal and limestone and heat them in a reverberatory furnace to produce raw soda.[2] The innovation of the Leblanc process in comparison with other processes, which also passed through sodium sulfate and transformed it into carbonate by utilizing charcoal, was the addition of limestone in the proportions (1:1:½). This made large-scale production possible. A third stage was added to refine the raw soda, which contained only 34–43 percent sodium carbonate: filtration, washing, evaporation, crystallization. Depending on the process used, one obtained "salt of soda," "black salt," "crystals of soda," or "caustic soda." Starting from two mineral substances, sea salt and sulfuric acid, what was then called a "mineral alkali" was produced through an ordered series of chemical reactions. The innovation resided in the very nature of the transformations used. Leblanc did not "extract" the soda, he made "synthetic soda." The Leblanc process can rightly take the title of a seminal event: although Lavoisier is the reputed founder of experimental or laboratory chemistry, Leblanc is depicted by historians of technology as the first to make artificial products.

In the saga of chemists, however, the inventor is seated far behind the scholar on the podium of glory. After Lavoisier's death, his colleagues forged the image of the enlightened genius who was martyred to a blind political struggle; in contrast, Leblanc's collaborators, Jean Jérôme Dizé in particular, eclipsed his memory with their attempts to assert their own rights. And far from celebrating the inventor's genius, posterity has a tendency to minimize Leblanc's merits by stressing instead the importance of the revolutionary wars, blockades, and other events that prevented the import of raw materials. Two contradictory logics thus cohabit the history of chemistry: on the one hand, the political circumstances of revolution are seen as a brutal force that blocked the progress of science by executing Lavoisier; on the other, they seemed to act as a spur to technical innovation.

What lies behind this contradiction? Besides the divergence in the interpretations of events touching on revolutionary history, the contrast between the two stories of founders betrays a profound difficulty in thinking through the relationship between chemistry as a science and industrial chemistry.

First of all, Leblanc's life is still a matter of controversy. In the biography edited by Auguste Anastasi (1884), a great-grandson, Leblanc appears as a solitary inventor who was a victim of revolutionary decrees. According to Anastasi, the Committee for Public Safety deprived Leblanc of the prosperous factory that he had set up at Saint-Denis with the financial support of the Duc d'Orléans. The government had decided to make his process public—available to all factories in the nation—to meet war needs. Leblanc was stripped of his rights as an inventor, his nationalized factory, the support of his guillotined protector. He was betrayed by colleagues, ruined by competitors, and finally reduced to suicide after years of vain attempts at restitution from successive governments.

This was the history that Leblanc related during his years of complaint and protest and that was embellished by his biographer (Anastasi, 1884). Contemporary historians suggest a less pathetic version of events (Gillispie, 1957; Smith, 1979). For one thing, the Saint-Denis factory, renamed Franciade, was not brutally closed by decree, since its production was already nearly stopped by the summer of 1793. It seems that it never reached full production for lack of sulfuric acid or financial means. Also, Leblanc seems to have been given consideration and was rather well treated by his colleagues on the Committee of Public Safety, who tried to find solutions acceptable to his interests and those of the Republic. The factory was given back to him in 1800, but he was not capable of managing it. Finally, and especially, the Leblanc process was neither the only nor the first to produce soda from sea salt, and its superiority was not evident to the eyes of his contemporaries, especially the consumers, who complained about the odor of his soda (which was due to the presence of hydrogen sulfide). The report of the commission mandated by the government in June 1794 to suggest ways of obtaining soda, while stressing the interest of the Leblanc process, extolled a whole range of techniques usable in different local circumstances. The Leblanc process became really competitive with natural soda only because taxes on salt were abolished in 1807 and import barriers were continued.

These two contrasting versions of the story—legend of a martyred hero or portrait of an inventor demystified by a professional historian— agree on one point: the lack of connection between this story and that of the chemical revolution. Although they were contemporaries and united by a tragic destiny, Lavoisier and Leblanc seem to have started two separate histories. Neither the invention nor the exploitation of this "revolutionary" process resulted from the chemical "revolution." Neither Leblanc nor his successors was able to explain why limestone was useful. Certainly the progress of chemical science at the beginning of the nineteenth century allowed them at least to determine the exact compositions of limestone and soda, the original and resultant products, but the process itself could only be understood when chemical reactions were studied at the end of the nineteenth century, when the last Leblanc soda works had closed. Like Minerva's owl, academic science takes flight at nightfall.

The gap between industrial history and the history of science is symbolized by the very name of the product in question. It was one of the first manifestations of the divorce between popular and scholarly language, because chemists call soda sodium carbonate (Na_2CO_3), and the word *soda* is reserved for sodium hydroxide (NaOH), which druggists and industrialists call "caustic soda." Some chemist-entrepreneurs, like Chaptal, wanting to keep close ties between the laboratory and the factory, attempted to speak the double language, but the split widened in spite of the numerous books on chemistry published during the nineteenth century.

Around the Soda Works

Finally and above all, the original feature of the Leblanc process was that it built up around the soda works a real technological system. It was the beginning of a complex industrial development whose logic was far removed from the history of academic chemistry. It was this process, adopted in all countries and continually improved, that would set the pace of heavy chemical industry for nearly a century. At the very beginning of the nineteenth century, the first client of Saint-Denis was the Saint-Gobain glass works, which consumed three-quarters of the artificial soda in 1806. But other industries gradually grafted themselves onto the cycle of reactions in the Leblanc process. From the four raw materials used—sea salt, sulfuric acid, charcoal, and limestone—the sulfur was recovered as hydrogen sulfide, which could be reused to prepare sulfuric acid; and at the end of the first phase, an interesting product, hydrochloric acid, was obtained. This cycle fostered the integration and mechanization of the chemical industries. Upstream of the soda works, a sulfuric-acid works was installed. And later on, downstream, to make use of the hydrochloric acid that was otherwise given off into the atmosphere, to the peril of the inhabitants, the cattle, and the harvests, factories for making bleaching products, which were much in demand in the textile industry, were built.

These two offshoots also underwent technological innovations at the end of the eighteenth century. On the one hand, the production of sulfuric acid was scaled up following the introduction of the "lead chamber" by John Roebuck in 1746. The lead chamber was an enormous, enclosed tank of several cubic meters in which sulfur was burned with saltpeter and from which, after condensation in a water bath at the bottom, was removed an acidulated water, which was concentrated to produce sulfuric acid. Although the two phenomena that took place in the chamber—combustion and gaseous release—corresponded to the centers of interest of theoretical chemistry, at the end of the eighteenth century the lead chamber still kept its mystery. It had long been in use when Clément and Desormes advanced a theoretical interpretation of the reactions that took place in it.[3] In the immediate future this theoretical interpretation did not inspire any spectacular technological improvement, but it did have a considerable scientific impact, because it described for the first time what would be called a catalytic action.[4]

For the bleaching industry the pattern was reversed, since in this case the impulse came from the academics. In 1774 Scheele—who had been the first to isolate oxygen, which he did while studying the activity of various reagents, including that of muriatic acid on manganese—discovered and characterized a new gas, which he called "dephlogisticated marine acid." Ten years later, Berthollet undertook a series of experiments on this new gas. He renamed it "oxygenated muriatic acid," because that was the time of his conversion to Lavoisier's ideas, and published a study on its color-removing properties in 1786. The drapers of Rouen, who wanted to change the slow pace of bleaching on the grass but also to avoid the dangers of acid, were very interested.

But since the new gas was dangerous to manipulate and to transport, research on less noxious products continued. A mixture of marine acid (hydrochloric) and alcohol, Antoine Baumé suggested; a bleaching liquid produced by dissolving oxygenated muriatic gas in a solution of caustic potash, proposed Léonard Alban, a large manufacturer of acids and mineral salts, whose plant was founded in 1776 at Javel, near Paris. Under the name "Javel water" the liquid was manufactured, marketed, and distributed widely in the 1790s. But research continued, because Javel water remained expensive. Textile manufacturers in Scotland and

Lancashire preferred a product made by Charles Macintosh and patented in 1797 by the firm of Charles Tennant: "bleaching powder." This product would assure the prosperity of the English textile industry by reducing the time needed to bleach cotton to one week by around 1830. In France, even though the balance sheet looked very favorable after Chaptal's report on French industries in 1819, and encouraging political measures had been adopted, such as the exemption from the tax on salt in 1809 and the protection of the market by a ban on imports, the technique of bleaching with chlorine came into use more slowly than it had in England.

The effects of the emperor's politics are more visible in the beet sugar industry. The case is somewhat similar to that of synthetic soda, since the research took place well before the blockade. As early as the beginning of the eighteenth century, Olivier de Serres had pointed out the presence of sugar in beets. After intensive work led by Franz Karl Achard in Germany from 1777 to 1796, a first experimental factory was opened in 1796 in Steinau-am-Oder. It was a conclusive success. Other factories started up in Germany and Bohemia. Napoleon, in turn, encouraged this new sector and put Chaptal, Anselme Payen, and some entrepreneurs to work. In 1812 Benjamin Delessert opened a small sugar works in Passy, and Chaptal founded another one on his estate of Chanteloup. But it was especially in the north of France that sugar works would proliferate and form an integrated network of agriculture and industry.

Such was the niche in which the Kulhmann enterprise grew. Very modest when it started in Lille in 1825, it produced sulfuric acid with two lead chambers. The following year it added a Leblanc soda works to sell soda to the textile works in Lille, Roubaix, Tourcoing, and occasionally hydrochloric acid to manure producers. But since the culture of the sugar beet was a growing market there, Kulhmann seized the moment and constructed a fertilizer factory to produce superphosphates. With the opening of the railroad into Belgium a few years later, he expanded his market into that country. To Kulhmann this integrated development suggested innovations that which would increase the recovery from the sugar works: to rid the beet juice of the acids it contained, he used "animal black" and chalk. In 1834 Kulhmann retrieved waste hydrochloric acid from the soda works to regenerate the animal black; then in 1838 he proposed a massive utilization of chalk to form "calcium sucrate," which was eliminated with all the impurities. Later on, other improvements—from technologies as varied as setting up boilers for evaporation and the selection of improved varieties of sugar beets suggested by Louis de Vilmorin in 1856—assured the prosperity of the sugar works and related industries.

The interdependence of the interests of the technologies and governmental policy was essential for revving up the chemical industries in France, but these interests interfered with those of a third player: academics in industry. Although we do not for the moment have enough facts to appreciate the phenomenon as a whole, it seems that in certain industrial sectors, where the application of the Leblanc process led to a broadening and diversification of chemical products, the entrepreneurs called upon chemists to invent or perfect their systems of production. Thus the Saint-Gobain Company, created in the seventeenth century, called upon Gay-Lussac. Starting as an advisor visiting the factories and supervising production, Gay-Lussac suggested an innovation in the fabrication of sulfuric acid: a tower to recover the nitrous gases that escaped from the lead chamber, which would be used to oxidize the sulfur dioxide to obtain sulfuric acid. He developed the idea for nearly ten years, with the director of the Chauny factory. This process, known by the name of "Gay-Lussac's tower" and put on-line in 1842, proved to be very profitable, because it used fewer nitrates. Gay-Lussac, who had become a member of the board of directors in 1839, became its president in 1844. Although he held a powerful position, Gay-Lussac had to give up all his patent rights to the company, which had a monopoly on production in France, and negotiated the patent with Charles Tennant in England.

In this alliance between academic scientists and industrialists, the academics contributed much more than their scientific expertise, because their sphere of intervention was not limited to the factory. The government, in fact, seemed to respond to pressure from scientists and industrialists on political decisions. This occurred several times: Baron Dupin and then Gay-Lussac demanded the abolition of taxes on the salt used for soda making; Gay-Lussac and Thénard made a plea in the Chamber for the assimilation of industrial property with intellectual property. As for the law on child labor, it was championed by the Industrial Society of Mulhouse, which had reduced the work day for children to eight or ten hours and wanted this measure to be adopted widely in order to keep competition fair. In his capacity as a peer of

France, Gay-Lussac intervened in the discussion of this projected law, which was approved in 1841.[5]

A technological network of interlocking processes and integrated industries, a human network of political alliances and combined interests—such were the supports for the fledgling French chemical industry at the beginning of the nineteenth century. In any case, after 1850 it was no longer France but England that was the primary producer of Leblanc soda. Lancashire and Tyneside were dotted with soda works. Small, tranquil villages had become foul industrial burroughs crowded with workers. This development may be traced to England's industrial needs: first textiles, glass, and soap, then paper and fertilizers consumed large quantities of acid and bleaching powder. To the internal demand, which grew incessantly, foreign markets must be added, thanks to progress in transportation on the one hand and tarif reduction on the other. Difficulties with supply and fluctuations in the price of sulfur imported from Sicily caused the English to look for new ways of making sulfuric acid and to use pyrites from Ireland and Norway. Legislation also contributed to the industrial boom around the Leblanc process. In 1863 the Alkali Act obliged soda works to recycle 95 percent of their hydrochloric acid. Several processes (Weldon in 1866, Deacon in 1868) were put into operation to produce the chlorine for bleaching powder by transforming hydrochloric acid with the help of a catalyst and the action of the oxygen in the air. In spite of these efforts, English industry could not use all the hydrochloric acid produced by its soda works. In the 1860s commercial treaties lowered the price of British soda by 15 percent.[6] This signaled the end for the soda works in the south of France: there were twenty-two in Marseilles in 1830; seven, in 1880. These small plants could no longer compete: they lacked capital for modernization and had to add the cost of transportation to their selling price, because most of their markets were located in the north and the east.

Obstinate Inventors

Here, to parallel Franco-English rivalries, is the sad but true story of an unfortunate inventor (Veillerette, 1987). Philippe Lebon (1767–1804), a young engineer who had learned about Lavoisier's work on hydrogen and was passionately interested in fire machines, had the idea of using the gas that escaped from the combustion of wood for heating and lighting. After repeated experimental tests, he applied for several patents in 1799 and 1801 for "thermolamps or stoves that heat cheaply and offer several precious products, a power source applicable to any kind of machine." Dreaming of lighting the city with his gas, Lebon offered his invention and his apparatus to the government. No answer. Confident and determined to win out, Lebon tried a gigantic publicity stunt. He rented a private house right in the middle of Paris and illuminated it in a grandiose style: the rooms, antechambers, salons, façade, park, and grotto at the end of the garden were regularly lighted up during several months of 1801. In spite of the success of this spectacle and his experimental demonstrations, performed in the presence of Guyton de Morveau, Fourcroy, and Chaptal, Lebon could not bring his product to market. Manufacturers did not believe in it and individuals scorned it. Clément and Desormes considered Lebon's invention a joke, and Charles Nodier used his pen and his talent to ridicule such eccentricities. Having squandered the entire fortune inherited from his family on his experiments and illuminations, Philippe Lebon died at the age of thirty-seven—ruined, bitter, and disappointed. You Englishmen, you be the first to light up!

On the other side of the Channel, projects for lighting with gas seemed less fantastic. In the country of charcoal, William Murdoch proposed to burn coal rather than wood charcoal. In 1798 he started lighting the Bolton and Watt foundry with gas experimentally. And in 1805 gas lighting was finally installed in the building, as well as in Philips and Lee's cotton spinning mill. With the enthusiastic collaboration of a German industrialist, Winsor, Murdoch founded the Chartered Gas Light and Coke Co. in 1810. The director, Christian Friedrich Accum (1769–1838), already famous for his private school of chemistry, spread the new technology through his book, *Treatise on Gas Lighting,* published in 1815. During this period Winsor accomplished Lebon's dream in Paris. He lighted the Passage des Panoramas and the Palais-Royal and then, in 1819, the Place du Carrousel. In the 1820s several societies were created, and these merged in 1855 into the Compagnie parisienne d'éclairage et de chauffage par le gaz.

Then, on the fringes of big cities, gas factories proliferated, and they ejected their highly polluting residues, ammonia waters, and coal tar into the streams. The Gas Light and Coke Co. did construct a distillery to produce coatings for roofing and cordage with the coal tar and the pitch, but all the Navy Office's ships and all the roofs in England were not enough to use so much pitch. And what to do about the dirty water, full of ammonia, draining directly into the Thames? In the columns of the *Gas Gazette,* little notices offering ammonia residues to the highest bidder appeared regularly.

Many thought of using these residues to make soda. The reaction, which followed Berthollet's laws, had been known since the beginning of the century.[7] But there was a long way between knowing the reaction and putting it into practice, especially in the period of full production of the Leblanc soda works. James Muspratt and then Henry Deacon in England made the attempt but gave it up because of certain technical difficulties. In France a chemist, Jean-Jacques Théophile Schloesing, and an engineer, E. Rolland, made soda from ammonia for four years (1854–1858) in a factory near Puteaux, but they also gave it up, because the tax on salt increased the cost of the process considerably.

On this process, which had gobbled up the capital of so many shrewd chemists, the son of a Belgian industrial family would build a fortune. For reasons of health, the young Ernest Solvay (1838–1922) had hardly studied at the university or frequented the great laboratories in Germany or elsewhere. But he knew Berthollet's laws and the treatment of salts, because his father directed a salt works. Moreover, he was aware of the waste of ammonia waters in his uncle's gas company, where he had worked since he was twenty-one. He had the same bright idea as about fifteen chemists before him, and he took out a patent in 1861. Full of confidence because he was unaware of the blighted hopes of his predecessors, he left the gas company and set up a small experimental plant with his brother Alfred near Brussels. In 1863, encouraged by the first tests and with financial aid from their family, the two brothers constructed a soda factory at Couillet, near Charleroi. When it opened in 1865, nothing worked right. Incidents accumulated: the lime absorbed moisture and crumbled, the temperature had to be regulated constantly, and finally the carbonator blew up. The firm was liquidated. In July 1866, Ernest Solvay started two new ventures. At each step of the process he encountered a technical difficulty that he had to solve before going on to the next step. But far from becoming discouraged, he never lost confidence and went stubbornly ahead. The carbonation of the ammonia brine posed two important difficulties, which were resolved in 1868. To get a good absorption of the ammonia gas by the sodium chloride in solution, there had to be a good liquid-gas contact, and the heat given off by the absorption had to be dissipated. Ernest Solvay invented a cooling tower with a hundred plates and many sifters and baffles to increase the contact between the reagents. He also introduced a coke furnace to recover the ammonia waters as well as the by-products of coal tar distillation.

Solvay progressed from problems to solutions, and he had the foresight to patent every stage of the process in several countries as soon as it was developed. In December 1872, he was producing twelve tons per day. He constructed a soda works at Dombasle near Nancy, and Ludwig Mond, a former student of Bunsen, licensed his patents and introduced them into Great Britain, where he built a factory in Cheshire. Still, the Solvay process did not conquer English soil easily. The Leblanc soda works, prosperous and well entrenched, resisted for a long while: forty-eight of them joined together in 1890 and, as United Alkali, led the competition for several more years. It decayed only under the combined assault of the Solvay process and that of the soda produced by electrolysis from 1886 on. In France, where the Leblanc soda works were already declining, the Solvay process quickly won a dominant position. In 1874, 17,000 tons of ammonia soda were produced in France and only 56,000 tons of Leblanc soda. In 1905 the Leblanc process yielded only 6,000 tons, versus 270,000 tons by the Solvay process (Haber, 1958, p. 110). Germany and the United States, which had few Leblanc works, adopted the Solvay process directly and all the more avidly because internal demand was increasing with the rise of the organic chemical industries in Germany. In total, the Solvay process furnished two million tons, 90 percent of world production, in 1913.

While he was becoming a millionaire, Ernest Solvay was speculating on the great enigmas of the universe, for this lucky industrialist was tormented by the thirst for understanding. Impassioned by the origins and structure of the universe, haunted by the desire to unify its forces, Ernest Solvay devoted part of his time to physicochemical reflections. He undertook experiments and published several memoirs on heat, the unity of matter and energy, and finally a treatise, "On Gravity." In order to test and debate his daring hypotheses on gravity, Solvay invited a number of eminent physicists and chemists from various countries to Brussels in 1911. Without dawdling over Solvay's speculations, these eminent scholars discussed the problem that occupied them: "the theory of radiation and quanta." Thus the Solvay conferences were born, the second flourishing result of Solvay's entrepreneurship, and this time purely intellectual and philanthropic. The Solvay conferences on physics and chemistry, which have taken place periodically since 1911, played a major role in physics between the wars. As for Ernest Solvay, diverted from his ambition to be a physicist, he dreamed up other great projects in physiology and sociology (D'Or [undated]; Mehra, 1975).

Heavy chemical industry developed on the basis of a small number of relatively simple products and processes, which nevertheless involved complex and dynamic technological systems. Each product called for another: for raw materials and for industrial outlets in order to recover the by-products; very heavy equipment was required for soda production. But as many adjacent plants were built onto the main factory, networks multiplied the soda production and provided the incentive for expansion.

The Challenges of Nitrogen

That a peaceful, practically inert substance as generously spread throughout the atmosphere as nitrogen could occupy the energies of chemists for more than fifty years is one of those surprises that the properties of matter hold in store. That an element called *azote* (French for "not conducive to life") in 1787 could become vital to the effort to feed the galloping population of England, obsessed by the fear of famine,[8] haunted by Malthusianism, is a second surprise. This turn of affairs resulted from analyses made by Martin Heinrich Klaproth and Nicolas Vauquelin on samples of guano brought back from Peru by Alexander von Humboldt, as well as from later systematic analyses. Plants did not live on oxygen and light exclusively, as had been thought since Joseph Priestley, but also on nitrogen, phosphorous, calcium, and potassium.

That minerals played an essential role in the growth of living things was a finding that defied the old division of life into kingdoms and the long-held belief in a vital force. Not only did this discovery complicate the relationships between organic and inorganic chemistry, it raised a host of questions, which formed the core of a new discipline, agricultural chemistry. Where did these elements come from: the air, the water, the earth? Where did they enter the plant: through the roots, the leaves?

What path did they follow, what transformations made them assimilable by living tissues?

At the beginning of the nineteenth century it appeared that agricultural chemistry would be limited to explaining the ancestral practices of fertilization—rotation planting, the uses of lime, ash, bone powder, guano—by means of analysis. But the modest task of reorganizing agricultural practices scientifically hardly fit the conquering spirit of nineteenth-century chemists, especially not the all-powerful Justus von Liebig.

Chemists in the Fields

Liebig's ideas found the most receptive soil in England. The English, anxious to increase their agricultural yields, imported tons of guano and nitrates from Chile and Peru and produced phosphates and superphosphates from them. In the 1860s, Great Britain consumed 500,000 tons of fertilizer—as much as all of Europe and ten times as much as France.

Even so, the interest of the English in agricultural chemistry did not spring up suddenly, out of the pressure of need. As early as 1813 Humphry Davy devoted a whole work, *Agricultural Chemistry,* to these problems. To the question how mineral substances act within the plant, Davy answered: by means of humus, the only assimilable substance. He believed the activity of the mineral on the living to take place through a substance intermediate between the two kingdoms. The sole function of mineral elements, among which Davy counted nitrogen, was to stimulate the organic matter contained in the humus. Davy's work encouraged the use of guano in England, and more broadly it steered research toward humus, that buffer substance between the mineral and the organic. Detecting the fatty acids in the humus, analyzing them, and improving methods of analysis were the main tasks accomplished by 1840.

On this foundation a fertilizer industry began to develop. John Benett Lawes ran an experimental plant from 1839 to 1842: he spread bone meal treated with concentrated sulfuric acid on turnip fields. In 1842 he was granted a patent on a process for manufacturing "superphosphate of lime" and constructed a factory at Deptford, which marketed the product in 1843. Later, superphosphates would be made from mineral phosphates, and then England would be less favored than France, which had access to deposits of them in North Africa. But the process of treatment with sulfuric acid, which was indispensable to making the phosphate assimilable by plants, remained the one that had been developed in England around 1840.

In 1840 Liebig, already recognized as the head of two successful research schools, entered the field of agriculture (Rossiter, 1975). Far from pigeonholing the chemist in the role of nature's assistant, the farmer's adviser, Liebig secured for chemists the role of expert, judge, and master of agricultural methods. This was the program outlined in his *Chemistry in Its Application to Agriculture and Physiology,* published in German in 1840, immediately translated into several languages and distributed widely, with nine German editions and nine in foreign languages (Munday, 1991; Finlay, 1991). In what way did Liebig reopen the question of agriculture, and why was he considered the "pope" of agriculture? The book discusses rather theoretical questions, especially concerning the relationship between the mineral and the vegetable. In this respect, Liebig was opposed to the dominant theory of plant nutrition by humus, and he insisted, on the contrary, on the role of the mineral elements. At the end of a long series of systematic analyses, he resolutely concluded that no evidence corroborated the role proposed for humus, and that the nitrogen fixed in plants came from atmospheric air, which was assimilated directly by leaves, or by roots, in the absence of leaves. Humus provided only carbonic acid before the appearance of the leaves.

How does a widely popular book emerge from such a theory? First, the success of the work had to do with Liebig's style: at the ends of chapters he did not hesitate to dispense practical counsel to farmers. In the main, this advice was based on a principle of balance that recalled Lavoisier. Applying balance sheets of reactions to the transformations that took place in the animal and vegetable world, Liebig turned the inputs and outputs into equations: everything taken out of the soil in the form of harvests had to be given back to it in the form of fertilizer or mineral elements. Let us recover the urine from the stables: one pound of urine wasted means 60 pounds of wheat lost. A second aspect helped make this book extremely popular: its polemical style. It opened by praising struggle in science—to avoid it would be to wipe out science, Liebig wrote, for controversy was the surest road to truth—and it effectively inspired as many lively polemics in agricultural circles as in chemical ones.

Jean-Baptiste Boussingault (1802–1887), professor of agricultural chemistry at the Conservatoire des Arts et Métiers, also stressed the importance of nitrogen in the growth of plants. And just after Liebig's work first appeared, he published a little book with Dumas covering exactly the same problems (1842). Later he openly contested Liebig's hypothesis on the fixation of atmospheric nitrogen. With Georges Ville he did a series of systematic analyses of rainwater to test his hypothesis. The results showed that only one kilogram of ammonia per hectare per year was provided by rainwater, which was far below the actual consumption. After Boussingault, Sir Joseph Henry Gilbert (1817–1903) of England, a former student at Giessen, used his analytical know-how to refute Liebig's theory. From a whole series of tests on the soil in fields he concluded that the nitrogen used came from the fertilizer incorporated into the soil.

But neither Boussingault nor any of Liebig's other detractors ever attained his popularity. In spite of the refutation of his central thesis, in spite of the failure to put it into practice, his "agricultural chemistry," as it was called, had a tremendous historical impact. On the one hand, Liebig's error was fruitful, for the polemic set off by it sustained a program of research for thirty years. Chemists and agriculturalists were both interested in nitrogen. Paradoxically, considering his emphasis on the nitrogen in the air, Liebig turned chemists' eyes to the soil. On the other hand, especially after the corrections and revisions introduced in the seventh edition in the 1860s, the book delivered a broad message that went beyond the specific question of the paths of nitrogen in plants: the farmer could not do without the services of the chemist. At the end of a hundred analyses of various soils, Liebig concluded that each species of plant needed its own kind of soil. Farmers and agriculturalists had only to look at the results! Liebig, unlike Davy, did not give them experimental protocols with which to do their own analyses. He recommended that they seek the advice of a professional chemist, who would help them to avoid impoverishing the earth, the source of their fortune. In sum, only chemistry could prevent the disasters of famine and the exhaustion of soils—it was her historic mission.

If the assistance of a chemist who has gone beyond the analytical exercises of the laboratory is necessary for a rational agriculture, it can also lead to disastrous results. Liebig had his own disappointing experience on English soil. In 1845 he teamed up with James Muspratt, a former student at Giessen whose father owned a small fertilizer factory near Liverpool. He took out a patent for six different fertilizers adapted to six different kinds of cultivation and began to market them in the hope of eventually replacing guano. It was a fiasco. The fertilizer solution formed a hard crust on the surface of the fields, which filled Liebig with bitterness on the subject of English soil. This anecdote illustrates a basic problem: chemists called out into the fields know only the universe of the laboratory, in which all extraneous circumstances are eliminated in order to control the reactions, whereas agriculture takes place under the open sky and must deal with all circumstances. The chemist cannot always excuse his failures by pretending they are the fault of the sun, the wind, or the rain.

To bridge the gap between a world subject to the hazards of meteors and the aseptic world of the laboratory governed by the laws of chemistry was the second task that was incumbent on agricultural chemistry. In the 1850s experimental stations were created in which chemists did full-scale experiments. Many were opened in the United States, where young chemists educated in Germany launched agricultural research after returning home and receiving university chairs. In England, in his Rothamsted station, Lawes did systematic tests with ammonia waters from the gas factories. The central problem, which necessitated years of research, hypotheses, and tests, was still the fixation of nitrogen by plants. A little progress was made in France, even though agricultural research was underdeveloped there and not much encouraged by the protectionism that taxed imported products at 25 percent. A Collège de France experimental station was created at Meudon for and by Berthelot. There he performed hundreds of analyses and demonstrated that clay soils and the microorganisms in them could slowly fix free atmospheric nitrogen, and the conditions in which this took place (Jacques, 1987, pp. 163–174). This was a different phenomenon from the direct fixation of atmospheric nitrogen by legumes,[9] which was demonstrated by the Germans. But Liebig was no longer in that field. After having left Giessen in 1852 for a chair in Munich, he devoted himself to private research. As for German agriculture, it flourished after the discovery, in 1856, of the mines of Stassfurt in Saxony, which made Germany the world's leading producer of potassium.

Factories with Two Faces

Where could the nitrogen needed for making fertilizer be found? All the industrialized nations found themselves competing to solve this problem. The race drove chemists to the factory to perform experiments involving high temperatures and pressures.

Recovering nitrogen from the ammonium sulfate in gas or coke factories was a convenient but limited method. Another solution was discovered at the very moment when Liebig formulated the problem: the deposits of saltpeter in the desert regions of Peru and Chile contained soda nitrates. Even after the development of processes to treat them, however, the Chilean nitrates were not the ideal solution: there was the problem of dependence on a monopoly and the risk of exhaustion.

The industrialized countries were all the more anxious to find more reliable sources of nitrogen because it had two uses in their factories. Since it produced greater agricultural yields, it was a key element in the struggle for life. Because it was involved in the manufacture of explosives, it was a key element in the death struggle for military superiority.

In fact, it was largely in the works of mining and civil engineering—tunnels, canals, bridges, and viaducts—that explosives were used at the end of the nineteenth century. The nitroglycerine made by Asciano Sobrero of Italy was too unstable to be produced wholesale and marketed. Alfred Nobel (1833–1896), third son of Emmanuel Nobel, an entrepreneur in explosives, met Sobrero while he was in Paris working under Théophile Jules Pelouze in 1850. He began to produce nitroglycerine in the family factory in Stockholm. During a violent explosion there in September 1864, his younger brother and four assistants were killed. In 1866 Nobel succeeded in stabilizing nitroglycerine, by absorbing it in a diatomaceous earth. The new explosive, dynamite, was sold in the form of a powder with a detonator based on fulminate of mercury. Production went from eleven tons per year in 1867 to 1,350 tons in 1872. Nobel was assured of revenues from his thirteen factories in Europe plus two in the United States. In 1875 he put a new explosive on the market: blasting gelatine, a gum made of a mixture of nitroglycerine and nitrocellulose. In 1887 he introduced another, exclusively for military use. Alfred Nobel, who controlled the market by creating two multinational holding companies, one in Paris and the other in London, had thus accumulated a pretty fortune at the end of his life. It gave him the opportunity to leave 32 million Swedish crowns in his will to endow a foundation that awards five annual prizes to encourage the advancement of science and to promote the cause of peace. The renown of the Nobel Prize must not blind us to the fact that with Nobel began a new era of military art, in which the old black powder would be left behind for new nitrate explosives, such as the melinite (1886) with a base of trinitrophenol used by the French army and the TNT (trinitrotoluene) that would be made in Germany.

The strategic importance of fertilizer factories and the large sums invested in research on nitrogen made extracting nitrogen directly from the air an obvious solution, since nitrogen constitutes 80 percent of the atmosphere. But the development of a process to achieve this goal was extremely costly. The process consists of forming nitrogen oxide (NO) by a strongly endothermic reaction, which requires temperatures on the order of 3,000°C, obtained by an electric arc, and even so the return was only five percent. The principal obstacle was that during cooling the inverse reaction (exothermic) took place; also, it was necessary for the cooling to take place quickly, so ovens capable of producing high temperatures in a very constant and localized way had to be constructed. Next, nitric acid had to be produced from the oxide obtained in the previous step and combined with lime to make the calcium nitrate usable as fertilizer. The first attempts were made by William Crookes in a station near Manchester in 1892. In 1903 Christian Birkeland and Samuel Eyde patented a furnace that required electric power on the order of 4,000 kilowatts and thus required a very low cost source of electric energy. Only the Norwegian Society of Nitrogen, which could use waterfalls to produce electricity, managed to industrialize the process, around 1910, at the cost of very heavy equipment.

Germany had bet on another process, based on calcium cyanamide, which used a litle less electricity. It consisted of fixing nitrogen on calcium carbide by heating to 1,000°C: $CaC_2 + N_2 \rightarrow CN_2Ca + C$. The calcium cyanamide could then be hydrolyzed with superheated vapors to obtain ammonia. The first factory to exploit the process was founded in 1910 at Knapsack, near Cologne, and the second at Trotsberg, near Munich. The process was bought by the English to construct a factory in Norway, as well as by an American, Franck Washburn, who founded the American Cyanamid Company and set up a factory near Niagara Falls.

Another ambition filled chemists' thoughts at the end of the nineteenth century: to achieve the direct synthesis of ammonia from its constituent elements. In 1888, after having formulated a law of the displacement of chemical equilibrium, Henry Le Chatelier (1850–1936) defined the conditions of temperature and pressure that would theoretically be necessary to perform this synthesis in the presence of a catalyst: if the process was run at 500–600°C, the combination of hydrogen and nitrogen remained incomplete, but since the reaction was accompanied by a decrease in volume, it should be possible to increase the yield by increasing the pressure. Le Chatelier received a patent in 1903, but he abandoned industrial development after an explosion. The problem of synthesizing ammonia was twofold: it was necessary to master not only the science of chemical equilibria but high-pressure techniques as well.

Germany accomplished this in 1913 with a team of three specialists. The first, Fritz Haber (1868–1934), was a chemist schooled by Bunsen and Hofmann and then at the Technische Hochschule in Berlin; he taught organic chemistry at Karlsruhe while working in his father's business at the same time. In 1908 he entered the Badische Anilin und Soda Fabrik (BASF) and began to work on ammonia. In 1909 he achieved a first synthesis with an osmium catalyst in a laboratory reactor at a pressure near 200 atmospheres and 550°C. The key idea of the Haber process was to recycle the gases at high pressure to increase the yield. Immediately Heinrich von Brunk, director of BASF, assigned two engineers to the project: Carl Bosch (1874–1940), an expert in metallurgy whose task was to find materials resistant to pressure and corrosion, and Aldwin Mittasch, an expert in catalysis to work with Haber. In 1910 BASF bet on the results: it decided to give up the Norwegian process of the electric arc and invested considerable capital in the construction of a pilot factory at Oppau, near Ludwigshafen. In 1912 it produced one ton per day, but there were still technical problems with the catalyst. Mittasch systematically reviewed all possible catalysts in order to find the optimal one. In 1913 BASF went into industrial production with 8,700 tons per year. Haber was raised to the peerage and became wealthy, since he received one pfennig for every kilogram of ammonia sold. He obtained a chair at the Kaiser Wilhelm Institute in Berlin in 1912 and the Nobel Prize in 1918.

This success was one of the major results of the interaction between universities and industries, which Germany had been fostering since the 1870s. For the first time a fully integrated research project was undertaken, led by an interdisciplinary team, with a risk strategy and a large investment of capital. One can therefore see in it the origin of what is called "research and development" today.

The synthesis of ammonia was a founding event in other ways as well. It inaugurated a new paradigm for the chemical industry, one based on the mastery of high-pressure techniques and a dynamic approach to chemical kinetics (which used the idea of the rate of reaction, i.e., of yield, in spatio-temporal terms).

Finally, the synthesis of ammonia had a geopolitical dimension. BASF became the first production group for basic chemistry, and Ludwig-shafen the great center of world chemistry, because Haber's process necessitated elaborate integration. To begin with, an industrial infra-structure to furnish pure nitrogen and hydrogen was required. And since ammonia is difficult to transport, factories to manufacture explosives and fertilizers had to be installed nearby. On the eve of the First World War, when the supply of nitrates from Chile or elsewhere might be cut off, it was a decisive asset for Germany.

The Struggle for Dyes

Out of the thick, black coal tars, the unsavory by-products of the cokeries, would come mauve, fuchsia, magenta, Prussian blue, gold, acid green, indigo . . . all the colors that brightened the elegant clothing of La Belle Epoque, the Great Work of nineteenth-century chemists. Their particu-lar philosopher's stone was carbon, that singular atom capable of bond-ing with itself and of forming complicated, stable, and varied molecules.

Far from being a solitary quest, the making of synthetic dyestuffs was a collective enterprise that involved a number of groups and several generations of chemists. Because synthetic dyestuffs are one of the favor-ite examples of the emergence of science-based industries, the rise of the synthetic dye industry is usually described as though it were a direct application of pure science. Indeed, Hofmann's systematic research on coal tar distillation products supplied the raw materials, and the struc-tural theory of benzene played a significant role. It is certainly possible and convenient for historians of science to identify a speculative discov-ery, such as that of the benzene ring structure, as the turning point and precondition of all fine chemical industries.[10] It is undoubtedly pleasant for academic chemists to evoke the historical case of the development of synthetic dyes to reinforce the image of pure science generating useful technological innovations.

The case of the synthetic dyes industry, however, could just as well provide a good example of the distance between chemical theory and industrial practice, so long was the path from the laboratory discovery of coloring substances to efficient and profitable plant processes. Here, however, it will be presented as a good illustration of the complex interactions between academic chemistry and industrial chemistry.

Industrial Expertise

Emphasizing structural organic chemistry as a precondition of the manufacture of synthetic dyes in standard histories assumes a strong discontinuity between the first and second halves of the nineteenth century, between analytic and synthetic chemistry, between extractive industries and the manufacture of artificial, synthetic products. One of the main results of recent scholarship on the technology of dyes has been to question this tacit assumption from different perspectives.

It has been pointed out that early in the nineteenth century academic chemists were already collaborating with industrial firms, because of the booming textile industry's demand for laboratory studies of the constituents of the natural dyes extracted from plants. The textile chemists (then called "colorists") used to work on the spot—in a laboratory that traveled from firm to firm, developed in Great Britain, which held the leading position in the cotton industry—in Mulhouse, where André Koechlin and Dollfus founded the Société Industrielle de Mulhouse in 1825. Mulhouse became so famous for coloring and printing technologies that designers went there from all over Europe to create new fashions (Fox, 1984; Drouot, Rohmer, and Stoskopf, 1991). The Alsacian case is exemplary but not exceptional in the mid-nineteenth century. Large dyeing firms, such as Guinon, Marnas & Bonnet, and Renard Freres in Lyons and W. Spindler in Berlin, had already given up artisanal techniques and developed industrial processes with the help of colorists (Homburg, 1983).

In 1862 the French chemist Antoine Jérôme Balard, commenting on the common fabric of academic and industrial interests, remarked, "In a sense, the new factories are overgrown laboratories" (Balard, 1862, p. 213). In fact, the shift from artisanal to industrial, standardized processes did not occur as a direct result of laboratory research. In a detailed study of dyes technologies, W. J. Hornix (1992) stresses the difference between the laboratory and industry. The industrial equipment used in the synthetic dye industry did not originate from the laboratory bench.

It is impossible to consider processes requiring reactors with steam jackets or pressure filters as scaled-up laboratory apparatus. The equipment was in fact derived from other industries, such as steam power and extractive industries, especially the extraction, purification, and fixing processes used in making natural dyes and calico printing.

Moreover, the relationship between science and technology has never been one-way. Anthony Travis (1993) argues convincingly that the dye industry had a tremendous impact on the development of organic chemistry. The dye industry produced the materials needed by academic chemists, who were often consultants to industry, for their investigations. Whereas Hofmann's work is used as an example of academic chemistry providing raw materials to industrialists, Travis argues that Hofmann was able to identify aniline red, blue, and violet as members of the same family (in the terms of his type theory) thanks to samples of these colors provided by the Simpson, Maule and Nicholson plant. Thus, industry made a significant contribution to the achievement of one of the main goals of academic chemistry at that time, the classification of organic compounds. Although Kekulé's benzene theory was certainly not a direct outcome of an industry, industry helped prove the validity of the ring structure and provided legitimacy to structural chemistry.

Although the synthetic dye industry cannot be reduced to "applied organic chemistry," it must be admitted that advances in organic chemistry fostered the development of science-based industries. In the midst of a ferocious competition among rival industrial nations for market leadership, stimulated by a variety of patent laws, emerged a new profile for chemistry.

To describe this forty-year evolution, we adopt L. F. Haber's (1958) classification of three successive generations of synthetic colors—mauve, yellow, and indigo—because it allows a clear overview of the economic struggle between France, the United Kingdom, and Germany. The international rivalry was encouraged by a series of World Exhibitions that took place almost every four years between 1862 and 1900, allowing the public to compare the dyes displayed by various companies.

Mauve

William Henry Perkin (1838–1907), a young assistant to Hofmann at the Royal College of Chemistry in London, is generally celebrated as the first chemist to prepare a synthetic dye. The story is that while attempting to synthesize quinine, a very valuable substance for colonial expeditions, he discovered aniline purple. Adding four atoms of oxygen to two molecules of toluidine, he obtained an uninteresting red-brown precipitate. Repeating the experiment and replacing the toluidine with impure aniline, he obtained a black substance from which he extracted aniline purple.

Perkin patented his dye in August 1856. Confident of its commercial success, he did not hesitate to leave the Royal College of Chemistry in order to manufacture the product, in spite of Hofmann's warnings. Several trials allowed him to convince a manufacturer and to set up a factory at Greenford Green. Perkin quickly found the appropriate methods to perform the different transformations from coal tar to purple (see table below, from Haber, 1958, p. 82), and he created a mordant for dyeing not only silk but cotton and wool as well.

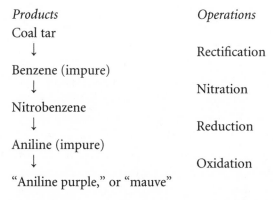

Products	Operations
Coal tar	
↓	Rectification
Benzene (impure)	
↓	Nitration
Nitrobenzene	
↓	Reduction
Aniline (impure)	
↓	Oxidation
"Aniline purple," or "mauve"	

Aniline purple made a great impression at the Paris International Exhibition of 1862, especially in the imperial court of Napoleon III. It was known in France as "mauve," because the Alsacian firm of Thann-Mulhouse was already manufactoring the aniline under that name. The other star of the Paris Exhibition was the fuchsia red (called *la fuchsine,* or magenta) manufactured by Renard Frères in Lyons. It had been discovered by Emmanuel Verguin, who prepared it by oxidizing toluidine mixed with aniline (Leprieur and Papon, 1979). Perkin's aniline purple stimulated intense chemical research on alternative methods of preparing dyes from aniline both in Britain and in France.

France seemed to be the leader in aniline colors because of its expertise in coal tar dyes. Perkin's famous purple was not, in fact, the first dye of this kind. Picric acid yellow had been manufactured by Guinon in Lyons since 1845 by treating phenol (or phenic acid), another heavy product of the distillation of coal tar. It is interesting to note that the French report on the Colors Section of the 1862 World Exhibition, written by J. F. Persoz, described the new artificial colors as a "revolution" whose effect would be to discredit the skill of dyers to the benefit of a purely commercial competition. In response to this challenge, Persoz also invited the government to promote new kinds of institutes of technology to sustain the efforts of France's industries in their competition.

The new dyes did, in fact, act as agents of discord, and much patent litigation took place among rival firms. The Société Renard et Frères believed it had secured the marked for *la fuchsine,* because this color was protected by a Verguin patent. But the same color could be prepared by other methods of oxidating aniline that were within the reach of everyone, for neither elaborate knowledge nor very intricate development was necessary. Moreover, one could make other dyes, such as Lyons blue, manufactured by the Société Monnet et Dury, or violet and green, invented by Hofmann and Perkin, from magenta. Renard et Frères then launched a judicial suit against its French and foreign competitors. Renard won the case for magenta, but it was a fragile victory, for Compagnie La Fuchsine, using a patent for which it had a monopoly, did not attempt to enlarge the market (Cayez, 1988, pp. 13–16). This reflected the inhibiting effect of French legislation: the 1844 law on patents protected the product but not its process of fabrication. It therefore discouraged research into improving processes or finding ways to use by-products. This flaw, relatively harmless in mechanical technology, was amplified in the chemical industry, because the same product could be synthesized a dozen different ways. Verguin and Renard's patent on magenta, which obliged any company to obtain a license to produce it, did not prevent the Gerbers in Mulhouse from developing a cheaper process using arsenic. They simply emigrated to Basel in order to be able to exploit it freely. In the thirty-nine German states, the patent legislation was so complicated that it left a lot of room in which to maneuver and allowed the same product to be manufactured by using different processes. The Paris 1867 World Exhibition, which presented eleven new shades to a dazzled public, marked the high point for French industry. But it was also its swan song: the Société Renard et Frères failed in 1868. Other French manufacturers, which had bet on other colors, survived, but France, weakened by the loss of Alsace and Lorraine, did not manage to make the cut for the second generation of dyes. Exit France.

Manchester Yellow

The second generation of dyes still used aniline as a base, but the synthesis process included a new step, "nitrosation"—the action of nitrous acid on aniline—that produced very stable dyes, which could be printed directly without using a mordant.

The operation that produced diazonium salts was known by 1858, thanks to a publication by Peter Griess of England, who was working, like Perkin, at the Royal College of Chemistry under Hofmann's direction. But it became interesting only after the synthesis of several aniline dyes. Heinrich Caro (1834–1911) of Germany, a cloth printer living in Manchester in 1859 and working in the firm of Roberts Dale and Co. manufacturing Perkin mauve, developed a new process. With Carl Martius he reacted diazonium salts with aromatic amines and in 1864 created two azo-dyes, which he named after their birthplace: Manchester yellow and Manchester brown.

With its infrastructure of textile industries, Manchester was becoming the capital of synthetic dyes around 1865. It attracted German chemists who wanted to start industrial ventures. Heinrich Caro was joined by Ivan Levinstein, who left Berlin to set up an enterprise that became one of the most prosperous. Azo-dyes gradually replaced the first aniline dyes. By 1902 they represented half the commercial dyes: 385 out of 681.

So it was England that seemed to be the winner, thanks to the success of azo-dyes. But there was a disquieting symptom in the 1880s—German chemists and industrialists were returning home, where they would exploit the English patents. In spite of its coal resources, financing, and the textile market that stimulated it, the British dye industry, which had been so prosperous and innovative in the 1860s, would be overtaken in the third-generation battle over dyes. A number of German firms, notably Bayer, would dominate from the 1880s onwards by taking advantage of two political measures taken by the state of Prussia. An 1877 law on patents that protected German inventions—processes rather than products, as in France—stimulated the search for new, more economical methods of synthesizing the same products and, at the same time, created a new market for knowledge of organic chemistry. Moreover, the creation of institutes to educate technicians, the Technische Hochschulen, and then in 1911 the Kaiser Wilhelm Gesellschaft Institut, directed by the ministry of education and supported by industrial firms, would encourage a very strong alliance between the state and private interests. Exit England.

The Madder Root Supplanted

In spite of their brightness and variety, aniline-based dyes had not supplanted the natural dyes of vegetable origin that were still made in Languedoc and Provence. But the first synthesis of a dye with an alizarin base in 1869 turned the whole economic landscape upside-down. Witness this story of the misadventures of Jean Henri Fabre, entomologist and professor at Avignon. One of the great resources of the region was madder, a herbaceous plant that furnished the famous madder red. As an amateur chemist, Fabre managed to extract the principal dyeing agent, alizarin, from it, and to purify and concentrate the chemical so that it could be used for direct printing on cloth. Since the process was not demanding and rather practical, he associated himself with a manufacturer to exploit it industrially. Persuaded that he would finally be able to increase his meager salary as a high school teacher with the revenues from alizarin, he turned down an attractive teaching position in Paris and decided to return to his dear Provence and the insects on the slopes of Mont Ventoux. But a cruel disappointment awaited him: "Hardly had the factory reached its stride when the news was heard . . . the tinctorial principle of madder had just been obtained artificially; a laboratory preparation turned the agriculture and industry of my region on its head. Having dabbled a bit in the problem of artificial alizarin myself, I know enough about it to predict that in the not so distant future work with a retort will replace that in the fields" (Fabre, 1925, p. 374).

Even though the retort did not yield alizarin dyes by itself, this anecdote shows that their industrial synthesis was predicted, carefully planned, and the fruit of a long research program. The competitors were Perkin in London and Carl Graebe (1841–1927) in Berlin, in collaboration with Carl Theodor Liebermann (1842–1914). Graebe and Liebermann worked in an institute that would soon become, with the support of the Badische Anilin und Soda Fabrik (BASF), the famous Technische Hochschule (Haber, 1958; Beer, 1959).

The starting point for alizarin synthesis was another product of the distillation of coal tar, anthracene, but the operations were much more complicated than for aniline.

Anthracene + Dichromate [Oxidation]

Anthraquinone + sulfuric acid [Sulfonation]

Anthraquinone monosulfonate + soda [Alkalinization]

Dioxyanthraquinone, or alizarin

The whole process conceived by Graebe and Liebermann could not become commercially viable until fuming sulfuric acid was used, which lowered the price. This idea came from Heinrich Caro, who had gone back to his country at the end of the 1860s to join BASF. On June 25, 1869, Caro, Graebe, and Liebermann applied for a patent in England for the synthesis of alizarin. On June 26, Perkin applied for *his* patent. He was one day too late. The granting of the patent precipitated the fall of British dyes and the vertiginous ascent of Germany. German companies went into production immediately, lowered prices, and conquered most of the markets. BASF built an empire on the synthesis of alizarin, and the traditional dye industry was ruined. The French government tried to protect the farmers and entrepreneurs of Languedoc and Provence by erecting customs barriers and by using the army as a captive market. The main effect of this measure was that when French soldiers marched to the front in 1914, they were easily visible from a distance in splendid uniforms with madder red pants.

Indigo and Industrial Research

The decay of the vegetable dye industry, which appears to us inevitable today, came at the cost of very great efforts and heavy investments of capital. What would be called "Big Science" in the twentieth century began with artificial indigo. Its synthesis was so complicated that thirty years of assiduous research were needed to achieve a commercial product. Only businesses with a large market and big profits could enter the race for indigo. The financial investment was so large that they ran the risk of bankruptcy in case of failure. As a result, the attempt to synthesize indigo led to the formation of cartels.

Indigo involved a second innovation in chemical practice, called "programmed synthesis" by Meyer-Thurow (1982). The subordination of technology to the caprices of invention was over; the era of research and development had begun. Adolf von Baeyer (1835–1017) was the hero of this process, in which university and industry joined hands. A former pupil of Kekulé, in 1875 he took Liebig's chair in Munich, where he directed a laboratory instructing about five hundred students per semester and published hundreds of articles. Baeyer did the first work on indigo as soon as the structure of benzene was discovered in 1865. In 1880 he made his first synthesis with a derivative of cinnamic acid. The process was bought immediately by BASF and Höchst, who contracted to finance continuing research—and share the profits from it, as well. One hundred fifty-two patents on the synthesis of indigo from cinnamic acid were applied for, but indigo was obtained only in small quantities. In 1882 Baeyer proposed another process with a base of orthonitrotoluene, which was immediately tried by BASF and Höchst. But it used up so much toluene that it necessitated increasing the distillation of coal tar and therefore produced benzene and naphthalene in excess of other industries' needs for them. Research pursued ceaselessly finally resulted in industrial production in the 1890s. Several complex processes were developed, which used naphthalene and produced phthalic anhydride by oxidizing it with concentrated sulfuric acid. Synthetic indigo was marketed by BASF in 1897, but research continued and a new process by Carl Heumann and J. Pfleger, which was much cheaper, was used by Höchst from 1904 on. Victory for Germany on two fronts, academic and commercial: Baeyer received the Nobel Prize for chemistry in 1905, and by 1910 no natural indigo could be found on the European market.

With alizarin-based dyes began not only an "industrialization of invention," in the form of programmed research with high investments, but also offensive commercial policies. The great German companies were not content to exhibit their products in world exhibitions. They advertised them by giving out samples, brochures informing consumers about the manufacturing conditions, and invitations to visit the factories. Mass production, the reduction of prices, and the stimulation of demand were the arms that conquered markets. On the eve of the First World War, Germany was producing 85 percent of dyes worldwide, while France, which had started off honorably in the era of synthetic dyes, produced only two percent.

This imbalance had repercussions well beyond the dye market. First, organic syntheses, which consumed sulfuric acid and coal tars, gave a boost to heavy industry and obliged it to adapt. Germany, which already possessed Solvay soda works, did not hesitate to invest in research on the electrolytic production of soda. The process developed in 1886 by the firm of Matthes and Weber of Duisbourg was the origin of the development of a powerful electrochemical industry in Germany. More broadly, the politics of innovation put to work on the synthesis of indigo was the origin of the research laboratory as an integral part of industrial factories (Homburg, 1992). Already favored by unplanned collaborations between universities and industry, the management policy of major Germans firms like Höchst and BASF laid the basis for a further expansion of the industrial research system. Industrial research labs employed a growing number of full-time chemists and gave them high salaries and managerial tasks, along with an interest in the profits of the firm. Research laboratories within the factories were diversified to allow research on several processes at once, along with systematic trials. Thanks to this integration of research, the firms that led the synthetic dye market also led the market for the new photographic and pharmaceutical products.

One example of the impact of industrial research is the Bayer Company at Eberfeld (Meyer-Thurow, 1982). When it was founded officially in 1863 under the name of Friedrich Bayer and Co., it was manufacturing a little magenta and some aniline dyes in Barmen. In 1872 Bayer moved to Eberfeld and began making alizarin on a large scale. Threatened by the drop in prices, the company hired Carl Duisberg, a chemist already employed at AGFA, in 1884 and created a research section in the factory. This turned out to be a major decision for the future of the business—for the chemical industry as a whole. In the beginning, the laboratory consisted simply of some benches separated by cubicles in which seven or eight chemists would do routine tests. But in 1889 Duisberg turned the whole laboratory toward research. A dozen full-time researchers were pursuing several projects. In 1900 the laboratory's founder became a member of the board of directors and convinced them to diversify the research labs; 144 chemists were working for Bayer.

One of the most spectacular effects of this new industrial organization was the reorientation of production toward pharmaceutical products. The success of the hygienist movement in the second half of the

nineteenth century had increased the demand for antiseptic products.
Since these were for the most part derivatives of the phenol family,
they were not completely foreign to the products of dye factories. By
the mid-nineteenth century, Britain seemed to be leading the pharmaceutical industry, since it was the first to use chloroform (in 1847)
and phenol as an antiseptic in surgery starting in 1867 with Lister. But
in the 1880s Bayer overtook it: Duisberg, leaving the indigo to BASF
and Höchst, specialized in the systematic study of the antipyretic properties of the intermediate products of dyes. In 1888 Bayer marketed
phenacetine; and in 1898, it brought out a pill that made a good profit,
acetylsalicylic acid, otherwise known as good old aspirin. Certainly the
Duisberg laboratory did not succeed this well in every area: there was
no positive result from the research on synthetic rubber or photographic materials, but even considering these failures, industrial research allowed Bayer to occupy the field and to block access to the
creation of new firms.

By winning the synthetic dye battle, Germany assured its future in all
areas of the chemical industry. When the twentieth century began,
France and England had already been crushed and humiliated by the
display of German power at the World Exhibition in 1900. However, the
French report on the chemical industry at the Paris Exhibition made
special mention of the American companies' displays, harbingers of an
emerging and promising chemical power.

The Race for Materials

If one wanted to characterize the twentieth century according to the type
of material that dominated its technology—as the Stone Age, the Bronze
Age, and the Iron Age earned their names—one would be in a quandary.
A number of candidates present themselves immediately: rubber, plastics, light metals and metallic alloys, composite materials, ceramics,
silicon. A jungle of diverse materials—this is our age.

The search for new materials is certainly not new, but it has been
particularly avid in this century marked by two world wars, not to
mention the Cold War, which stimulated research into replacement
products and increasingly sophisticated systems of defense. Moreover,
new industries—such as the automobile, aviation, aerospace, and elec-

tronics industries—called for specific materials with particular properties such as lightness, resistance to high pressures and temperatures, electrical and magnetic conductibility, a particular type of surface, and so on. A new rationale of production was the result: given a function or performance to achieve, find the material that has the needed properties. It was thus assumed that there was no existing material from which one could eventually make this or that product. In many cases the definition of the product preceded the conception of the material.

This motivation for research and development in the domain of materials transformed chemistry's scientific and industrial profiles profoundly. Although some industrial sectors remained completely separate, chemistry was increasingly called upon to serve other sciences and industries, to integrate itself into sectors such as transportation and electronics. These new tendencies were the end result of a long history, of which we will discuss only a few episodes—aluminum, rubber, plastics, and composite materials. These examples also illustrate some of the constraints that influenced a number of current research policies.

Aluminum

That a material is "new" does not necessarily mean that it is unknown in nature or the chemist's laboratory. The new material may have been known and identified for a long time, without having left a mark on the industrial landscape.[11]

This was the case for aluminum. Although this element is very widely distributed in nature, it did not find a place in industry until the twentieth century. Because in nature aluminum always occurs tightly bonded with oxygen, it was necessary to use very powerful means to separate it, and so the destiny of aluminum was linked to electricity.

The method of electrolysis that followed Volta's discovery of the battery was not powerful enough to isolate aluminum, in spite of repeated efforts by Humphry Davy and Christian Oersted. Friedrich Wöhler managed it in 1827, by using the activity of potassium on anhydrous aluminum chloride ($AlCl_3$). But achieving in a factory what had been done in a laboratory was quite another matter; it would require more than half a century of effort and testing. In 1854 the French chemist Henri Sainte-Claire Deville (1818–1881) invented a process in his laboratory at the Ecole Normale Superieur: he replaced potassium by the less expensive sodium and the AlCl₃ by a more easily separable double salt (NaCl·AlCl₃). The alumina needed for the preparation of aluminum chloride was provided by treating bauxite with carbonate of soda, using a high-temperature technique. It was in Gard, not far from bauxite deposits and near a Leblanc soda works, that the Compagnie des Produits Chimiques d'Alais, founded in 1855 by Henri Merle, produced aluminum for the first time.

This novelty arrived at the right time, just as a world exhibition was opening in Paris. In the rotunda of the exhibition's panorama, next to the Sèvres, the Gobelins, and the crown jewels gleamed a sumptuous table service for a hundred, ordered by the emperor from Christofle. Was it gold or heavy silver? Not at all; this symbol of richness and luxury displayed on the imperial table was made of an ordinary metal covered by a fine layer of aluminum by electrolysis. This process, which required a battery, was called "galvanoplasty" in honor of Luigi Galvani. It was developed in 1838 by Moritz Hermann von Jacobi, who used it to reproduce the imprint of designs engraved on copper. Two patents received in 1840 and 1841 were bought by Christofle, which hoped to revolutionize jewelry in this way. The emperor and his entourage, seizing upon any occasion to calm revolutionary ardor, then exalted chemistry, which was said to eliminate inequalities of class and wealth. "Henceforth, and this is vital," wrote Charles P. Magne in *L'Illustration* on November 17, 1855 (p. 331), "beautiful shapes, richly decorated patterns, finely engraved services, flatware, candelabras, bowls and plates adorned by the genius of statuary and ornamental art, can be made at affordable prices for modest fortunes and be seen elsewhere than on millionaires' tables."

In spite of the great expectations of the World Exhibition and the support of the emperor, everyday life seemed less brilliant for aluminum. Buyers were not knocking down the doors. The still too costly process made aluminum a precious metal. Affordability was delayed another thirty years, until a later Paris Exhibition, in 1881, would demonstrate electricity at an affordable price. It was only when the dynamo had replaced the battery and it was learned how to transport electrical energy that industrial electrochemistry was able to make real progress. Two young men, Charles Martin Hall (1863–1924) of the United States and Paul Héroult (1863–1914) of France, applied for a patent on a manufacturing process by electrolysis. The process envisioned by Héroult was described in his patent of April 23, 1886, under the title of *Procedé électrique pour la fabrication de l'aluminium:*

> It consists of placing the alumina in solution in a bath of cryolite, melted on one hand by means of an electrode in contact with the crucible of agglomerated charcoal that contains the cryolite, and on the other by another electrode of agglomerated charcoal, dipped into the bath like the first one. This combination produces the decomposition of the alumina by a low-voltage current. The oxygen goes to the anode, which burns at this contact; the aluminum is deposited on the walls of the crucible, which constitutes the cathode and precipitates the slag into the bottom of the crucible. The bath remains constant and continues to serve indefinitely if it is supplied with alumina. The positive electrode, i.e., the anode, must be replaced after the combustion, but this combustion prevents polarization and ensures in this very way the constancy of the energy and the action of the current.[12]

Immediately after applying for his patent, Héroult tried to convince the Compagnie de Produits Chimiques d'Alais (about to be renamed A. R. Pechiney and Co.) to test his process. That was his first failure. His second failure was at the Rothschild Bank, which preferred to invest in a competing process invented by Adolphe Minet. Finally, in December 1887, Héroult licensed his process to the Société Metallurgique, located near a waterfall on the Rhine at Neuhausen, before setting up a factory at Froges in Isère. During this time, Charles Hall, who received his patent in 1889, joined the Pittsburgh Reduction Co., which later became the Aluminum Company of America (Alcoa). The Pechiney Company, still lagging behind, was obliged to stop production of aluminum at Salindres. But in 1897, thanks to a site, the Maurienne valley, which offered an enormous hydroelectric potential as well as railroad transportation, Pechiney reentered the race and produced aluminum with Hall's process. To the first factory, Calypso, others would be added, until this valley of peasants and farmers was transformed into a great industrial center into which a seasonal labor force flocked in summer, when the water was flowing in abundance. At the end of the century, the world production was almost a million tons of aluminum by electrolysis at a production cost that went from 87.50 francs per kilogram in 1886 to 3.95 francs in 1895 (Aftalion, 1987, pp. 69–71). The production was so high that at the beginning of the twentieth century prices fell drastically—from 4 francs per kilogram in 1908 to 1.5 francs. This resulted in a perpetual search for technical improvements—increases in the size and strength of the vats and innovations in industrial refining processes. These improvements would allow light alloys to spread into daily life and to answer the needs of the rising aviation industry.

Electrothermal innovations generally accompany the development of electrochemical processes. Not only did Héroult conceive the idea of using electricity to dissociate salts by electrolysis and produce aluminum, but in 1887 he also invented a furnace that used electricity to produce high temperatures. Henri Moissan and T. L. Wilson of Canada used the electric furnace to produce calcium carbide in industrial quantity. At the beginning of the century this was a very interesting substance, because it could be used to produce acetylene, the primary basic material of the organic chemistry industries and the source of light for miners' lamps and vehicular headlights.

Electrochemistry and electrometallurgy created a style of industry with an original kind of research organization (Morel, 1991). The many improvements in the electrolysis process that allowed the price of aluminum to be cut were the work of the vat engineers, the men in the field who had to resolve practical problems at the crossroads of the disciplines of electricity, metallurgy, and chemistry. In this domain research could take place only at large scales, in the actual production vats. So each factory had its "secret garden," one or two vats for trying out various improvements (which were not necessarily communicated to other establishments). In the 1920s, after a crisis in the aluminum market and the hooking up of factories into a general network of electrical supply, new technical problems arose. Then new research departments were created, above and beyond the usual research laboratories, to work on methods of production principally devoted to electrolysis. This gradually transformed the structure of aluminum plants: information circulated more freely among the firms and research extended to longer-term improvements (Leroux-Calas, 1991).

But the institutionalization of a research laboratory, outside the factory, led to a competition *within* enterprises—between the "vat men," workers and production engineers or "electrolysists," and the research engineers, who had been vaguely disdained by the others until then. This hierarchical relationship would reverse itself in the second half of the twentieth century. At first in order to diversify production processes, specialists from the outside, primarily from academic institutions, were employed by firms. Finally, in the 1960s, the introduction of the micro-processor for controlling operations completely changed the organiza-tion of work. Research became focused increasingly narrowly, and researchers became more and more highly specialized. The time when workers would be allowed to spend time fiddling with the vat and production engineers would be seen as all-powerful was over. Factories produced, laboratories investigated. All at once aluminum—produced henceforth thanks to the improvement of refining processes, control of impurities, reductions in weight, and computer-aided research—could be considered a new material, if one goes by the economic criterion of a rate of growth higher than the average growth rate of industrialized countries (Cohendet, Ledoux, and Zuscovitch, 1987, pp. 14–15). But whatever definitions are used, the history of aluminum invites us to revise our concept of a new material.

Chemical Warfare

April 22, 1915: panic and rout in the front ranks of French and English troops asphyxiated by clouds of a suffocating gas, since named "Ypérite," in memory of a battlefield on the outskirts of Ypres. So began chemical warfare. In defiance of the Hague convention, poisons were unleashed on the front in the form of gas or white phosphorous shells. Even if the actual effectiveness of gases in combat was doubtful, since they were often dispersed by the wind or blown back on the troops who set them off, the psychological effect was decisive. The event caused a realization among the allies: chemistry was a weapon, mistress of the battlefield. France discovered its own weakness in industrial chemistry, its depend-ence on foreign sources. A 1917 report attributed this cruel dependence to chemistry's lack of social prestige and stressed that "the recruitment of chemists here takes place almost exclusively among the humble classes, hardworking but poor" (Grandmougin, 1917, p. 59).

But the war changed the situation radically. As A. N. Whitehead ar-gued at the 1916 session of the British Association for the Advancement of Science, "organization of thought" and of action was key. Chemists had to give up their independent, academic, international style of inves-tigation for a collectivized effort in close alliance with governmental authority (Haber, 1968). In all countries war research takes place on an emergency footing—without concern for optimization, in secret, without publication, by multidisciplinary teams. The gas war was characterized by the reciprocity of the offensive and the defensive: research on the making of chloride gases—Ypérite and phosgene—was inseparable from research on protection—masks, buffers, hoods, respirators, not to mention psychological research on the morale of troops. In France it was the mathematician Paul Painlevé, named minister of public instruction and inventions in October 1915, who organized the scientific mobilization, with Charles Mourreu responsible for chemistry.[13] In Great Britain, British and Australian chemists were mobilized and the chemical industry was transformed into an arsenal of munitions (Macleod, 1993). At the end of 1916, a ministry of scientific research was created to try to alleviate the shortages of food and dyes. In Germany, Fritz Haber was the providential man: his process furnished 45 percent of the nitric acid needed for making explosives and, in addition, Haber created an organization for war research with six working groups, one of which was chemistry, under governmental control, which continued after the war. In each belligerent country the government intervened directly in matters of production.

Since international trade was interrupted by the war, all the industrialized countries had to push their chemical industries in order to ensure autonomous supply. This resulted in a second consequence of the war, or rather of the return to peace: overproduction. Everywhere industry had excess supplies: a surplus of fertilizer from the reconversion of explosive factories to civilian use; a surplus of dyes—284,000 tons were produced in 1924, although consumption was less than 154,000 tons. A depression began in 1921. In Great Britain and all the European countries, the chemical sector remained in a state of chaos into the twenties. To meet this uncertain and delicate situation, governments pushed for concentration. Thus in Germany the dye industries, which had been associated since 1916, merged into IG Farben, leading producer of chemicals worldwide in 1925. In 1926 several British companies merged into the Imperial Chemical Industry (ICI). In France the Société Chimique des Usines du Rhone, based in Lyon, merged with the Société Wittmann et Poulenc Jeune, set up in Paris in 1928.

Even in the United States, where the federal government was less interventionist, the war contributed to the building of great empires, notably the Du Pont company. Starting before the entry of the United

States into the war in 1917, the chemical industry collaborated in the war effort through a consultative commission created by the federal government to provide explosives to the allies. The famous Weizmann process was used to obtain the acetone vital for making explosives and protective coatings for airplanes.[14] American companies, which had already achieved a remarkable increase in power from 1880 to 1914, thanks to the economic development of the country, prospered during the war and profited from the tariff protection granted by Congress in September 1916 to diversify their products—in particular, dyes which had previously been imported from Germany. So American industry was the true—perhaps the only—winner in this war.

Before changing the chemical landscape because of surpluses, the First World War shook the chemical industry because of shortages. From these shortages came the third consequence of the war, the search for replacement products. Whether it was butter, sugar, milk, coffee, or any other basic product that was in short supply, chemists were needed to find a substitute. The German term *ersatz* entered the English (and French) language because the Germans became masters in this area. They were the first to produce synthetic gas[15] and, above all, a product with a brilliant future: synthetic rubber.

Rubber

Strictly speaking, rubber was not a new material in 1916 (Le Bras, 1969). Already known to the Mayans, then brought to Europe by Charles Marie de La Condamine in the eighteenth century, rubber from *Hevea brasiliensis* trees remained a curiosity as long as it had no use other than erasing pencil marks from paper. It entered industry in the nineteenth century, when Charles Macintosh of Ireland used it to rainproof cloth. And it became a truly interesting material when Charles Goodyear (1800–1860) invented vulcanization: by adding sulfur and heating the mixture above the melting point, he obtained a rubber that was elastic and resistant to variations in temperature. Many uses for rubber were found, but still there was no large-scale production.

In 1845 a certain Robert William Thomson had registered a patent in several countries for an air mattress in an envelope of elastic rubber. A patent without a future, it seemed. But with the new craze for the bicycle and, a few years later, the automobile, that all changed. In the 1890s John

Boy Dunlop (1840–1921), an Irish veterinarian, reinvented the pneumatic tire to use on his son's bicycle, according to legend. Shortly afterward, Edouard Michelin (1859–1940) patented the first dismountable pneumatic tire for automobiles and went into large-scale production. The automobile industry increased the demand so much that rubber production rose quickly from 8,000 tons in 1870 to 94,000 tons in 1910, to more than a million tons in 1936.

With the increase in demand, Amazonia, with its great forests of heveas, became a strategic zone: some people made a fortune there, while the miserable inhabitants toiled at the harvest. In order to give the United Kingdom access to rubber, a brilliant subject of his majesty, H. A. Wickham, arranged to smuggle out some hevea seeds, which he planned to germinate at Kew Garden and then acclimatize them in Ceylon and Malaysia. The hevea plantations of Southeast Asia quickly challenged those of Brazil. By 1914 the production of cultivated rubber outstripped that of forest rubber harvested in Amazonia.

The intensive search for synthetic rubber was accelerated by World War I (Blackley, 1983, chap. 1). In the United States, two groups were working on the possibility of making rubber from 2,3-dimethyl-1,3-butadiene. In Europe, two patents were requested simultaneously for a substance synthesized from isoprene with metallic sodium as a catalyst, one by Matthews and Strange in Britain, the other by the Bayer Company in Germany. In 1915, Germany, cut off from its sources of supply by the blockade imposed on it by Britain, made a major effort to synthesize rubber. The first synthetic rubber, prepared by Bayer from dimethyl-butadiene, hardly survived the war, because its properties were too different from those of natural rubber and its price too high. Moreover, after the defeat, Germany was forced to turn over all its patents to the Allies by Article 297 of the Treaty of Versailles.

In the 1920s, the major advances occurred in the United States, thanks to the contributions of Ivan Ostromislensky, who used butadiene as the starting monomer; and Maximoff, who developed the polymerization reaction through emulsion. This process, which produced a polymer in suspension in an aqueous solution in the form of latex, resulted in a better rate of polymerization than the mass polymerization process that had previously been used and, moreover, allowed better control of the reaction. Also in the United States, in 1931, Du Pont made a synthetic rubber with very specific properties, Neoprene, obtained by polymerizing chloroprene (2-chloro-1,3-butadiene); and several years later a new family of synthetic rubbers appeared on the American market, produced by copolymerization of butadiene and acrylonitrile, which were brought to market under different names (Perbunan and Ameripol).[16]

In the period between the wars, however, Germany would win the battle for rubber by dominating all competitors. While in the United States all efforts were directed toward developing synthetic products with particular properties adapted to specific uses, in Germany the main goal was to find a polyvalent substitute for natural rubber. A research program launched in 1926 was encouraged by Hitler by means of a tax on imported natural rubber to finance research on synthetic rubber. At the Berlin auto show in 1936, IG Farben displayed Buna: a new kind of rubber obtained from butadiene *(Bu)* by a polymerization reaction catalyzed by metallic sodium *(Na)*. The properties of this product could be varied by copolymerization. The Germans put two varieties of Buna on the market in the thirties: Buna N (butadiene and acrylonitrile) had very specific uses; and Buna S, obtained by copolymerization of styrene and butadiene, was a solid rubber that had to be softened by heating or by adding plastifying oils—a process that lent itself, in fact, to practically all the uses of natural rubber. These two products made Germany powerful. In 1939, when the Second World War began, Germany synthesized 50,000 tons of rubber. The United States was making various commercial arrangements with Germany to obtain Buna S, which it did not produce itself, and even Buna N, when an explosion at Du Pont in 1937 interrupted its production of chloroprene. In 1941, when it entered the war, the United States produced only 41,000 tons of synthetic rubber; that same year, Germany raised its production to 120,000 tons.

World War II forced a radical change of U.S. policy in this area. The United States found itself cut off from all sources of supply: imports from Germany stopped in September 1939, and in 1942 the area of Southeast Asia in which its rubber plantations were located was occupied by Japan. There was no choice: it was imperative to produce a polyvalent rubber, an ersatz natural rubber, as Germany had done. The American Buna S version was called GR-S ("government rubber styrene"). The name difference was justified partly because the American product differed slightly from its German rival, thanks to an improvement in the polymerization process by emulsion, which helped to control plasticity. But the compound name indicated above all that this wartime rubber was obtained not only by the science of polymers but also by the will of the government. As early as June 28, 1940, President Roosevelt declared rubber a strategic and critical product. He created a mixed organization, the Rubber Reserve Company, composed of chemical industries and oil companies, to organize the production of synthetic rubber and protect the supply of all the basic products.

In 1942, the situation looked black: without a new source of supply of natural or synthetic rubber, America's reserve would be totally exhausted by the summer of 1943. On August 6, 1943, Roosevelt created a study commission, the Rubber Survey Committee, and gave full powers to its director. The result was spectacular. Production rose from 4,000 tons in 1942 to 180,000 tons in 1943 and 700,000 in 1944. Thanks to close cooperation between business laboratories and universities, and to rules for the communication of all technical information, the United States not only managed to solve the practical problems in polymerization by emulsion; it also made advances in the investigation of this *type* of reaction. After the war the results were published only in part, because rubber was still a strategic product, very sensitive during crises. Production rose again in 1951 during the war in Korea, and in 1959, during the Cold War, it rose above the 1944 record with 1,130,000 tons. The major profit of this policy went to American firms: Firestone, Goodyear and U.S. Rubber produced not only on American soil but throughout the world. Their success resulted from the fact that the rubber industry required a technical environment as well as an economic one. A good infrastructure of chemical industries was necessary to provide the ingredients of rubber—sulfur, zinc oxide, carbon black—as well as an infrastructure of industries to produce the textiles with which rubber was frequently associated. Moreover, large amounts of capital were needed to finance the heavy investment. Finally, and especially, consumers were needed—specifically, there had to be a market for automobiles with four tires.

So it was in the extreme situations created by two world wars that synthetic rubbers became essential. Unlike dyes, they did not replace the natural product right away. Between the wars there was a tendency to overproduce plantation rubber, which led to the lowering of prices. In 1934 an international regulation fixed an export quota for each producing country. In fact, synthetic and natural rubbers are different products. The synthetic consists of isoprenic links set end to end in the same direction, which form a long, wormlike molecule, while the natural forms a network spreading out in all directions with lateral chains. That is why the natural product remained master of the market for a long time. In 1950, plantation rubber still represented 75 percent of consumption (Reuben and Burstall, 1973, p. 27). Synthetic rubbers were used in small quantities because they were adapted to specific uses: mineral elastomers resist very high temperatures; hydrogel elastomers fill biomedical needs, such as supple contact lenses. Two processes for manufacturing stereospecific polyisoprenes developed in the United States in 1956 raised the hope that "natural synthetic rubber" would eventually replace the plantations of Southeast Asia. But these hopes were quickly dashed. Several factors favored natural rubber: the increase in the price of petroleum and its by-products; the fact that the hevea, unlike oil, is a renewable resource and does not pollute the environment very much. Together these factors assure a future for the hevea plantations.

Plastics

Plastics are an example of the triumph of a substitution product. Today they seem to us an almost natural class of materials because they are so familiar. They even seemed to take on an esthetic of their own in the 1950s. It was quite different in 1909, when Leo Hendrik Baekeland (1863–1944), a Belgian chemist who became an American, used the word *plastics* for the first time. The term then designated a larger class of products than it does today, including not only resins but also elastomers like rubber and synthetic fibers (Vène, 1976). The major characteristic of these first "plastics" was that they were conceived as "ersatz," replacement products for natural substances, which were either rare or costly.

The first synthetic plastic filled the need for a substitute for the ivory in billiard balls. A competition promised $10,000 to the inventor. A modest printer in the state of New York, J. W. Hyatt, found the solution in 1869, after six years of research: celluloid. Starting with cellulose, which is found in abundance in wood and cotton, treated with a sulfonitric mixture, he produced the nitrocelluloses until then used for explosives. He added another well-known natural product, camphor, an extract of the camphor tree, in the presence of a well-known solvent, alcohol, and a plastic substance resulted. Relatively elastic when cold, it could be blown up, drawn out, and welded when heated. Made into balls, stiff collars, barrettes—celluloid rapidly found a large number of industrial outlets by taking the place of ivory, shell, and horn.

Bakelite, prepared in 1907 by Baekeland from phenol and formalin, was a different type of plastic. When heated, it hardened, instead of becoming plastic as celluloid did. It inaugurated the line of thermohardeners. It was made not from natural materials but from by-products of industry: phenol was extracted from coal tar and formalin was made from the gas of coke furnaces. Since 1891, attempts had been made to manufacture resins to replace the natural gum-resin from phenol and formaldehyde. But it was only in 1907 that Baekeland mastered the pH, the respective amounts of phenol and formaldehyde, and the temperature to make the first entirely synthetic resin, bakelite. The General Bakelite Corporation, created in 1909, produced molds for a great many familiar objects in the 1920s and 1930s, in particular the first radios.

Although handicraft traditions had given way to modern industries, with polymer chemistry the time-honored practice of empirical know-how seemed to revive. Polymers were technological items before they were objects of knowledge. Early in the century polymers were often considered undesirable products: syrups impossible to crystallize or solids impossible to melt, they annoyed laboratory chemists, who were content to mention them in notes as "substances of unknown structure." After Baekeland, their thermohardening property was used to fashion a variety of objects, but their structure remained unknown. The hypothesis of molecules joined together in chains by ordinary interatomic bonds had been advanced as early as 1879 for isoprene, but most organic chemists around 1900 still thought that a pure body must be composed of identical molecules of small size. So the high molecular weight of the first polymers was explained as the effect of an aggregation of small molecules.

The investigation of the structure of polymers was primarily the work of the German chemist Hermann Staudinger (1881–1965), professor at the Zurich Polytechnical School from 1918 on. He introduced the notion of the macromolecule after studying polyoxymethylene, then polystyrene. Although his hypothesis was confirmed as early as the 1920s by crystallographic work with X rays by Hermann Mark, the idea of the macromolecule would take time to be accepted, so Staudinger did not receive the Nobel Prize until 1953. Since then, two processes for manufacturing polymers have been distinguished. Reactions of "polyaddition," or polymerization, which involve the juxtaposition of monomer

molecules, produce thermoplastic materials—such as the polyvinyl resins—that can be solidifed by heating and returned to fluid again by
reheating. "Polycondensation" reactions form macromolecules by
bonding two molecules together and eliminating a third: water, salt,
acid, alcohol, or amide. They produce thermohardening plastics that can
be molded and whose synthesis is completed during the manufacturing
process—such as phenolic resins like Bakelite, epoxy resins, and unsaturated polyesters.

The structures and reactions of polymers were thus investigated
within the framework of industrial research. As early as the 1920s and
1930s, IG Farben devoted five to ten percent of its profits to research on
polymers, and it led the race for plastic materials in 1939. After the
discovery of new paints in 1927, Du Pont also started a program of
research on polymers, by a team directed by Wallace H. Carothers
(1896–1937), and provided with ample funding. While developing a poly-
chloroprene rubber in 1931, Carothers also devoted himself to the study
of polyaddition reactions complementary to Staudinger's studies of
polycondensation in Zurich. During a study of aliphatic polyesters, one
of Carothers's collaborators, J. B. Hill, discovered that polyesters could
be drawn out into long fibers and be made considerably longer after
cooling. Even better, this cold stretching made them more solid and
more elastic. Unfortunately, these polyesters, having a very low melting
point and too great a solubility in water, were not really interesting from
a commercial point of view. From that time on, Du Pont directed
Carothers to work ceaselessly to find the right chain. After many trials
and errors, in which Du Pont invested 26 million dollars, he produced
polyamide 6.6, which had good, strong, elastic fibers that were insoluble
in water and had a high melting point (260°C). In 1936 an experimental
lot of stockings of a completely new type was produced. Industrial
production started up in a factory at Seaford, Delaware, in 1939. But
nylon would have to wait until the end of the war to invade the world.[17]
Wallace Carothers would no longer be there to savor the victory. As if
he were following in the footsteps of the distant ancestor of the chemical
industries, Nicolas Leblanc, he committed suicide on April 29, 1937—another example of the sad destiny of lucky inventors who put their mark
on their era.

In 1945, as after the First World War, the chemical industries that had
worked flat out to assure uninterrupted supplies of war materiel and had

carried on research in all directions found themselves overequipped and in search of markets. Then plastic articles started appearing all over the world. They gradually replaced traditional materials, such as steel, glass, and wood, everywhere. The "plastic years" inspired this critical reflection by a chemist in 1956: "It appears that many of the plastics spread profusely even into the marketplaces of the smallest African villages, are only pale imitations of admirable natural substances: ivory, bone, leather, wood, horn, etc., and their success is in large part due to the profound and unconscious tendency of man to bypass nature and admire himself through his own creations in the center of a pantheon of ersatz . . . All this rather feverish industrial activity, most often prodded by the black angel of war, reminds me of the 'nouveau riche' who blindly dispenses his capital without thinking of his heirs" (Baranger, 1956, p. 204). Though the warning against a frenetic and unreasonable consumption of fossil fuels—coal and oil—is still timely, the suspicion of a self-glorification of man in a "religion of the ersatz" seems very dated. Far from divorcing humanity from nature, far from causing "rootlessness" in the sense of Simone Weil, the use of plastics and synthetic furs spares the lives of a multitude of animals previously exploited and sacrificed to human needs. For the disdain of nature feared by the chemists of the plastic years, we should perhaps substitute the image of chemistry as protector of nature.

Materials on Demand

In the 1970s, after decades of mass production and a period of economic growth and inexpensive energy, there arose a new orientation toward research on specific characteristics and specific materials. Plastic materials were diversified, geared to specifically targeted performance, and designed as a function of the final product. This new trend manifested itself not only in plastics but in materials as a whole. After having achieved mass production and standardization, chemistry was put to work on a "made-to-order" civilization. In some cases the tendency toward the diversified production of made-to-order materials led to the actual reconception of industrial objects (Cohendet, Ledoux, and Zuscovitch, 1987). "All plastic" solutions or the dream of the completely ceramic motor still maintained the autonomy of the sector of production and even strengthened know-how in that area, but the same could not be said of composite materials.

These materials, made of a polymer matrix in a fibrous structure (glass, carbon, etc.) or a sandwich structure of three elements, imply by their very nature that various sectors of production work together to design a single product. The first class of materials conceived and developed by the applications sectors happened to be produced by the aerospace industry, which needed light structures capable of withstanding high temperatures. These first composite materials developed in the 1960s and 1970s were completely original in that they were the unique solution to an otherwise unsolvable technical problem. At the time, cost mattered little. Production, although it necessitated advanced technologies, remained artisanal. In the early 1980s, composites were beginning to be used in civil aviation and in sports equipment, but they remained expensive. Gradually, however, the technology was extended to materials of average performance and lower cost. Extensive use of composites led to the reorganization of the whole production sector, because close associations among industrial partners were needed.

What ambitions inspired such acrobatics? The main goal seemed to be to develop integration on all levels: integrating production by reducing the number of steps; reducing the number of pieces—three pieces instead of twenty-four, in the case of the Citroen BX's bumper—in order to decrease assembly costs; finally and above all, integrating the maximum number of functions into a material which, by this very fact, was revealed as more and more complex.

This functional dimension was even more pronounced in other materials used or sought to fill a specific function, such as transporting electricity or information, catalyzing chemical reactions, or, even better, in the case of biomaterials, performing functions vital for life. Contrary to other new materials that might be called "structural," these "functional" materials did not come to redefine a traditional sector of industry, such as automobiles or aeronautics, but *influenced* an entire sector of technology, in particular electronics and telecommunications. The most striking example is silicon, the basic element in integrated circuits. Monocrystalline silicon's physical properties offered a well-adapted solution to a crucial problem of the 1960s: the miniaturization of integrated circuits. The whole development of microelectronics came about because this problem was solved.

Would silicon become the single element around which a whole technological system would be built, as had happened during the expansion of chemical industries in the nineteenth century? It is true that silicon is, in a sense, a strategic element, but that does not mean that it is unique or irreplaceable. Far from being exclusive, silicon technology required a host of other materials, which affected several levels of computer science.[18] Moreover, the properties of silicon are now creating limits on the speed of *handling* information (the switching time of silicon transistors cannot be less than 10^{-9} seconds). Finally, silicon is not very well adapted for the *transmission* of information. In this area gallium arsenide is more interesting.[19] The linkage between telecommunications and electronics therefore motivates the search for other semiconductors.

Whether functional or structural, new materials are no longer intended to replace traditional materials. They are made to solve specific problems, and for this reason they embody a different notion of matter. Instead of imposing a shape on the mass of material, one develops an "informed material," in the sense that the material structure becomes richer and richer in information. Accomplishing this requires a detailed comprehension of the microscopic structure of materials, because it is in playing with these molecular, atomic, and even subatomic structures that one can invent materials adapted to industrial demands and control the factors needed for their reproduction, whether they are new or traditional.

The more the microscopic knowledge of material operates on the industrial level, the more the knowledge and know-how of the chemist are needed. The chemist is called upon to play an essential role everywhere in the field of materials. His task is to bridge the definition of a product's performance on the macroscopic scale with a definite microscopic structure and to develop synthetic and molding processes. Chemistry is no longer an independent actor of production with its own rules, as it was when plastics were developed. It now acts as an invaluable servant. More and more frequently channeled into different industrial areas, the expertise of the chemist is nevertheless everywhere in evidence, indispensable to a civilization that demands its materials "à la carte."

5

DISMEMBERING
A TERRITORY

A History for Chemistry?

Students learning how to "balance" stoichiometric equations today are working in a direct continuation of the "chemistry of professors." The heritage of the nineteenth century—the atomic masses as organized in Mendeleev's table, Gay-Lussac's law of volumes, the definite proportions of Proust and Dalton, the distinction between atoms and molecules—though created amid controversy and polemics, is now ensconced in the routine equations that represent the quantitative balance of reactions.

As for apprentice organic chemists, they study the population of reactants that are their tools in a direct extension of the chemistry of substitution. When, for example, the symbols R–X, R–COOH or R–OH appear, they mean that the reactant used leaves the identity of a radical, R, unchanged and changes the atomic grouping indicated. And the aspiring chemist learns to use all the properties of molecular structure that, since Kekulé and van't Hoff, synthetic chemistry has never stopped exploring, in order to piece chains of syntheses together. Decomposition, substitution, and synthesis are all reactions undistinguished by any fundamental property, because any combination can be described as both a decomposition and a substitution. The differences arise from functional contexts, each reaction being characterized by its practical end and corresponding to a relevant representation of the molecule.

The Belle Epoque

The nostalgic idea that "in the seventies" chemists had marked out a well-differentiated, coherent territory they could finally call their own, produced within their tradition, defined by the end of their controversies, is no doubt a "recurrent vision," one of those judgments about the past from the standpoint of the present. Pointing to a time when chemists finally spoke the same language, which would remain their common reference even through the fragmentation of their specialties, assumes that all chemists were, at that past moment, effectively agreed on their history as well as on their future. As we shall see, this was not the case.

The nostalgic idea that chemistry could once claim a territory that was later dismembered, that it found its fulfillment only to lose it, is nevertheless an interesting one. It points out a remarkable fact about the controversies that marked the end of the nineteenth century: these controversies still involved, as had been the case since the "origins," the question of chemistry's identity. In other words, "chemistry" had the status of a subject; its destiny was in the hands of chemists; it was a cause and a focus of dispute. It is in this sense, and not in the sense of a definable identity, that chemistry could be assigned a territory: it had its indigenous inhabitants and its borders, which could be crossed without consequences. So the 1870s celebrate not the memory of a moment of consensus among the participants, but a point of equilibrium—identifiable only from a distance—between an uncompleted past and a future that was already beginning. We recognize retrospectively the questions chemists were still wondering about as questions whose answers will not come from chemistry.

During the course of the nineteenth century a peculiarity defined the style of chemists. With a few exceptions—such as Dalton, who invoked the precedent of Kepler's laws "explained" by Newton—chemists insisted on a *practical* conception of the laws of nature: laws were the chemist's instruments and were connected to his practice. Some chemists, like Berthelot, prided themselves on this close connection: in chemistry, the exemplary positivist science, there was a clear difference between science and metaphysics, a difference that physicists sometimes forgot. Chemistry's greatness consisted precisely in its *not* transcending the facts learned in its practice. If the Newtonian dream of a deductive chemistry was not dead, those who alluded to it, like Dumas, only emphasized the distance between this dream and the modest, but solid, reality.

From this point of view, the "chemists' atom" as the constituent of molecular structures was based on a style that epistemologists would define as rational and lucid. As we saw, Kekulé did not "believe in" the atom; he needed atoms to think about the atomicity of elements. In other words, atoms gave end-of-the-century chemistry its terminology, but they could justly be suspected of being only a language, a useful fiction. They were not "discovered" as forces were discovered by Newton, alternating current by Hertz, or America by Columbus.

As a daughter of speculative alchemy with irreproachable morals, the professors' chemistry illustrates the austere morality of rational progress. But its epistemologically exemplary character also comes from the fact that, since Berthollet's Newtonian theory of the mixt and Berzelius's electrochemical dualism, no theoretical interpretation could be identified as the organizing point of view to which all the others must submit. There was no theory of the chemical atom: its stability and identity were as precarious as the agreement between the instrumental points of view that define it, each for itself. Chemistry had the structure of a network, not that of a tree: there was no consensus on what could be considered the trunk, the fundamentals from which branches might sprout.

That is why it is difficult to tell which of the chemists of the second half of the nineteenth century believed that atoms really existed, and which took them for fictions whose pretension to reality was by definition temporary and relative to their ability to organize the facts. Certainly, at the turn of the century belief in the reality of atoms was making progress, especially with organic chemists: Baeyer's model of molecules (balls that could be articulated by sticks), which was based on the hypothesis that the length of each type of chemical bond and the angles between the bonds were "atomic" (in other words, they could be described independently of the molecular structure), spread into the laboratories of organic chemistry. But it was still possible to oppose "atomic chemistry."

The Question of the Future

In 1910 many specialists in inorganic chemistry still thought that the atomic and molecular hypothesis was only a fiction and criticized the way those unobservable entities were presented as if they really existed. Although molecular structures had had an impact on organic chemistry, they had remained relatively peripheral in inorganic chemistry, which was more concerned with the variety of elements that entered into compounds than with the structures built by molecules.

But if the atom provoked skepticism from inorganic chemists, it was the target of much more radical questioning from two renowned physical chemists, Pierre Duhem (1861–1916) and Wilhelm Ostwald (1853–1932). To the idea of a science of the architecture of matter, Duhem and Ostwald opposed a counter-theory, that of the "energetics" of chemical transformations.[1] In order to marshall support for this possibility, both Duhem and Ostwald reread the past and made themselves historians of chemistry.

The history that, for us, leads to the atom and the molecule is told by Pierre Duhem in *Le Mixte et la combinaison chimique* (1902). It is a remarkable work in that it concludes with the possibility and the necessity of resisting the apparently irresistible belief in the existence of atoms. From equivalents and the corresponding raw chemical formulas to isomers and stereochemistry, Duhem insisted, neither an atom nor the assemblage of atoms that constituted a molecule had ever allowed a chemist to bypass experimentation. It was the discovery of new types of operations involving new distinctions among compound bodies that was the motive power leading the chemist to invent richer and richer representations of the bodies involved in these operations. "The symbols that modern chemistry uses—raw formulas, structural formulas, stereochemical formulas—are precious instruments of classification and discovery as long as one regards them only as elements of a language, of a notation suitable for visualizing in a particularly precise and arresting way the notions of analogous compounds, of bodies derived one from another, of optic isomers. When one wants, on the contrary, to look at them as a reflection, as an outline of the structure of the molecule, of the arrangement of atoms among themselves, of the shape of each one of them, one quickly runs into insoluble contradictions" (Duhem, [1902] 1985, pp. 138–139).

Even the law of multiple proportions did not escape Duhem's criticism: "It can neither be verified nor contradicted by the experimental method; it escapes the grasp of this method" (ibid.). Never indeed would experiment prove that the weights of two constituents are in the ratio of two whole numbers. And what about the idea of characterizing atoms by an attribute as arbitrarily variable as their valence! The atom with its valences is a useful and fertile instrument for classification, but it does not have the power to govern the phenomena that it arranges.

Likewise, in 1906 Wilhelm Ostwald shaped a critical history of chemistry that outlined its triumph as the science of molecular structure, but only to stress the precariousness of this triumph. The realism of stereochemical structures should not "resist facts," such as the following one discovered by the chemist Paul Walden: by substituting one atom for another in an optical isomer and then doing the inverse substitution, one could end up with the isomer of the opposite configuration[2] (i.e., of the opposite optical activity). How, Ostwald asked ([1906] 1909, pp. 150–151), could a simple substitution transform the optical properties of the molecule? For Ostwald the implication was clear: the structure had to be characterized as a "whole," capable of global transformation. The stereochemical structure, conceived from the spatial arrangement of its parts, marked the end of an epoque. The theories of the future would not be stated in these terms.

Ostwald proposed a description that anticipated Thomas Kuhn's paradigmatic revolutions or Imre Lakatos's characterization of the fate of research programs: "First a theory is developed to represent the variety of existing combinations by modification of a certain schema . . . But science keeps growing; sooner or later a disagreement between the actual multiplicity of observed facts and the artificial multiplicity of the theory is necessarily produced. Most of the time one first tries to bend the facts, if the theory, the possibilities of which it is much easier to embrace at a glance, cannot give up anything more. But facts are more resistant than all the theories, or, at least, than the men who defend them. And so it becomes necessary to enlarge the old doctrine appropriately or to replace it with better-adapted new ideas" (p. 147). And Ostwald announced that the period of "reciprocal adaptation" between the facts and stereochemistry was coming to an end. A radical reform was necessary, which should be centered no longer on static structures but on the intriguing subject of catalysis and on the relationship (which remained to be explained) between the relative stability of compounds and their internal energy.

There is, however, one point on which the more or less realistic atomists and the antiatomists such as Duhem and Ostwald would have agreed: the history of chemistry would remain in the hands of chemists, including physical chemists, who understood that their science could not be reduced to the "clear principles" of mechanistic physics. Because *we* know that this conviction was soon to be refuted, we can be nostalgic for a period in which chemistry *finally* resembled our chemistry while *still* being defined by its own history. Ten years after the publication of Duhem's book, chemistry would be given a "trunk" at last, but a trunk that emerged from elsewhere: from physics—even worse, from a physics that, at first atomist, would become "mechanistic," revolutionarily joining the great lineage of Newtonian physics.

But before that, as early as the end of the nineteenth century, some controversies found resolution in a manner that was already announcing the "dismembering" of chemistry's territory in the sense used in this chapter: a process that, starting with one of the traditional problems of chemistry, transformed it into an operating and instrumental reference for a development to which it no longer held the key. Thus, while inorganic and organic chemistry had maintained contact—the results of the one reverberated in the other, and vice versa—"biochemistry" would provide an answer to the ancient question of the relationship between the chemical and the living. The new discipline gave the chemist a useful and demanding role, but one devoid of any major conceptual stake.

A Chemistry for the Living?

In the 1850s, when Berthelot was meditating upon his great work on the synthesis of organic compounds using only the compounds of inorganic chemistry, Louis Pasteur had already established the boundaries that he thought laboratory chemistry would never cross: the molecular asymmetry that characterizes certain natural organic products. Pasteur's demonstration became a classic. It involved the action of a new type of agent, *Penicillium glaucum,* a mold that proliferates on tartrate. The mold would do what humans could not do.

Fermentation and Catalysis

The crystallographer could separate the inversely symmetrical micro-crystals composing crystallized paratartrate, dissolve them, notice that each solution now had a rotatory power, one determining a rotation of the light-polarization plane to the right and the other to the left. So he could assume that crystalline asymmetry was a manifestation of the molecular asymmetry responsible for optical activity. This observation introduced the problem of life, because, Pasteur said, it often happened that artificial derivatives—laboratory products derived from optically active natural bodies—did not have optical activity. Were they, like paratartrate, "racemic mixtures" of active optical isomers? And if this were the case, couldn't it be used as a criterion for distinguishing between the chemical and the living: living organisms being capable of producing one isomer to the exclusion of the other, laboratory procedures producing mixtures of isomers? Entrance of the mold on stage. Pasteur demonstrated that although the molds enthusiastically consumed one of the paratartrate isomers that he had separated out, there was no fermentation of the second isomer. When Pasteur presented the molds with the racemic, optically inactive tartrate, they clinched the demonstration: as the proliferation took place, the rotatory power of the liquid increased; the optical evolution stopped at a maximum, corresponding to the end of the biological process. The mold had accomplished the same separation as the crystallographer (Pasteur, 1860).

After the work of van't Hoff, "molecular asymmetry" lost its mystery and optical opposites became a favorite instrument for studying structures and reactions in organic chemistry. Pasteur's demonstration nevertheless opened a new chapter in the controversy over the relationship between the chemical and the living. For the first time, living things had been put to use in a scientific demonstration in a role in which they were asked not just to survive (as in Priestley's tests on the breathability of air) but, rather, to perform a quasi-technical activity. Like the synthetic chemist, the living organism did therefore have to deal with chemical structures. And, superior to the chemist in this, it could synthesize a structure without producing its mirror image at the same time.

But are we still talking about living things? Since the eighteenth century, the technique of fermentation, which Stahl called *zymotechnia* in 1697, had been a recognized discipline, the basis of the German beer industry. But this technique could not enter into the debate over the relationships between chemical activities and those of living things. For Stahl, as for Liebig, a contemporary of Pasteur, fermentation was not, properly speaking, a biological process. There was a difference between processes that seemed to reflect a "vital" power of organization—the development and maintenance of bodies—and processes that arose spontaneously from the laws of chemistry: putrefaction, corruption, degradation, decay.

Of course, Cagniard de Latour, Theodor Schwann, F. Kützing, and P. J. F. Turpin had all realized between 1835 and 1837 that the alcoholic fermentation of beer produced a deposit made of living cells, from which they concluded that fermentation was a product of the activity of yeast. But for Liebig, causality went in the other direction: everyone knew that a tree that died and rotted was invaded by mushrooms, but it was clear that this growth was an accessory consequence of the rotting, not its cause.

If fermentation interested Liebig nevertheless, it was because it had figured prominently among the phenomena that Berzelius had related to a "catalytic force" in 1839. The catalytic force of a body, whether it was fermentation or the acid that turned starch to sugar, was manifested when its presence alone, and not by affinity for the other reagents, could cause a reaction. According to Berzelius, the catalyst awakened "latent affinities" among the other reagents, thus allowing reactions of which they would not otherwise be capable at that temperature. More than a half-century later, the physical chemist Ostwald would greatly appreciate the expression "latent affinity." "The expression of latent or sleeping affinities simply means that there are chemical states which are not equilibrium states, and which, in spite of that, do not change over time. In these systems the chemical reaction is set off, provoked by the presence of bodies that act catalytically; what happens leads, like all chemical phenomena, to a more complete satisfaction of the afffinities, i.e., to the achievement of a more stable equilibrium" (Ostwald, [1906] 1909, p. 278).

For a late-nineteenth-century physical chemist like Ostwald, the main thing was that the catalyst set off reactions that were possible in principle, but for Liebig in 1839, the "catalytic force" too closely resembled a vital force *making a difference in the laboratory*. Liebig was a vitalist, but he judged that the vital force would never be the object of positive, laboratory knowledge: the chemist could identify and re-create all the chemical transformations in living beings but could not reach the organized principle that forms these many transformations into one living being. That is why Liebig countered Berzelius's too-mysterious catalytic force with a hypothesis on the transmission of motion from the catalyzing body to the reagents. In the case of the fermentation of beer, he affirmed that yeast was a body in decomposition: the "atomic" vibration that accompanied this decomposition induced, by contact, the chemical transformation of fermentable bodies.

To Pasteur, ferments were living, organized bodies, and fermentation, far from being spontaneous decay, was an integral part of the chemistry of life. In 1857 he showed, contrary to Liebig, that the process of fermentation was largely independent of the nature of the fermentable body (the "nitrogenated material" could be a simple ammonia salt). On the other hand, it was totally dependent on the presence or absence of the ferment. Each type of ferment, *not* each fermentable milieu, produced a specific fermentation process. The ferment was therefore the cause. Fermentation—like decay and, as would soon be discovered, like disease—was determined by the activity of living cells; it could not be reduced to the spontaneous chemical processes of death.

Enzymes

Liebig would never be convinced, and indeed the situation was far from clear. What about the "diastase of germinated barley," a cellular extract capable of provoking fermentation? Pasteur distinguished between "ferments figurés," which were active to the extent that the cell was living, and "ferments non figurés," which could be separated from the organism and which Wilhelm Kühne would name "enzymes" in 1878. But wasn't this an artificial distinction, purely ad hoc?[3]

Moreover, the idea of connecting life not with specific chemical substances but with a "motion" was taken very seriously by the physiologists of the time (see Fruton, 1990). Was not life perpetual activity, whereas all laboratory reactions reached a final, inert equilibrium? Why not attribute dynamic properties, which would also resolve the enigma of the origin of life, to living matter, to the "protein substance"[4] that constituted the cellular milieu? Materialist thinkers, antivitalists, welcomed this definition of life as perpetual activity. In his *Dialectics of Nature* Engels wrote: "Life is the mode of existence of protein bodies, the essential element of which consists in *continual, metabolic interchange with the natural environment outside them,* and which ceases with metabolism, bringing about the decomposition of the protein."[5]

The biochemical analysis of living beings went full speed ahead. In the 1880s cellular protoplasm, which had been thought to be a homogeneous substance, "albumin," was revealed to consist of "proteins" and also phospholipids and "nucleins." In 1897 Eduard Buchner succeeded in extracting from yeast what he called "zymase," which was capable of producing fermentation. Therefore, it was possible to have fermentation without living cells. Besides, "enzymatic" activity was no longer a menace to the chemist but a model. Catalysis had become the favorite and indispensable tool of the synthetic chemist. Each catalyst offered a new possibility for synthesis: to be able to provoke a reaction was to be able to go from one molecule to another in a precise, reliable way, and with a good *yield*.[6] The "catalytic force" was the indispensable ally of the chemist, as it now seemed to be the ally of life, which was also "full of catalysis." Like the activity of living things, activity in the synthesis laboratory created means narrowly defined as instruments in the service of an end. All chemical transformations were controlled and specific there. Only the ends were different: biological in one case, economic in the other. As Ostwald noted precisely, the acceleration of slow reactions was important for the chemical industry, because "time is money" (Ostwald, [1906] 1909, p. 260).

The Victory of "Dead Molecules"

As early as the First World War, "zymotechnology" had been renamed "biotechnology." Molds, ferments, and bacteria were living things, and their activity opened a new field in the competition between man and nature. While organic synthesis produced artificial molecules by a costly and labor-intensive process, living things in the service of humanity promised marvels, such as complete recycling—for example, the evening newspaper converted to sugar for the next day's breakfast (see Bud, 1992)!

The dreams of biotechnology left the chemist only the humble role of the keeper of the Lavoisian balance sheet: identifying products and their entrances and exits from the black box, which henceforth was bacterial culture. For the scene inside the black box to be lighted up, biological catalytic action would have to lose its mystery. As long as it was only an empirical observation, it was perfectly compatible with a dynamic conception of life, based on a hypothesis of the instability specific to "living molecules" in a permanent state of flux. Baeyer's work on organic dyes and Emil Fischer's on sugars, purines, and, between 1899 and 1908, the structure of proteins (identification of the peptide bond and synthesis of the first artificial polypeptides) marked the beginning of a long-term research program, that of the analysis and synthesis of the various molecular structures that constituted living things. But the import of this program depended on whether or not the purifying activity of chemistry killed what it identified. Were "crystalloid" substances "dead" molecules, separated from the environment on which their "living" properties were dependent?

As early as 1898, however, Fischer had proposed what is for us the correct interpretation of enzymatic action, which fit easily into the chemistry of "crystalloids." It was the famous model of the key in the lock, in which the substrate and the enzyme have complementary forms, allowing the enzyme to fix the substrate. With Paul Ehrlich (1854–1915), this model made a hit among the immunologists. But those who believed in the colloid theory saw it only as a hypothesis pertaining to a strategic operation whose meaning was only too clear: submit living enzymes to the principles of stereochemistry.

Could a crystallized molecule have biological activity? If it could, the structures identified by biochemistry since Baeyer and Fischer were the true actors of life. Such was the terrain that the "antidynamic" biochemists would choose for a frontal attack on what appeared as a form of disguised vitalism in the twentieth century; at the end of the nineteenth century, it had been a question of "materialist" theory. In 1930 the American biochemist John Howard Northrop finally succeeded in crystallizing pepsin in conditions such that most of his adversaries were forced to recognize it was pure and still capable of enzymatic activity. But a quarter of a century later, the battle was still not over. The French biologist Jacques Monod (1910–1976) recounted the story: "I remember it was early in 1954, and I lectured in the States . . . on the subject that the interpretation of the so-called dynamic state of the proteins was wrong . . . You've got to realize that it raised an absolute *furor!* There was this Hegelian idea, you know, that this dynamic state was a sort of secret of life . . . At that time, the only people who were fully aware that this business of the dynamic state of protein molecules couldn't be right were the crystallographers. Because it couldn't be right if they got good crystals. To which, of course, the cell physiologists or biochemists said, 'But you are looking at *dead molecules!* " (cited in Judson, 1979, p. 391). For Monod it was a real crusade, because the transformation into Stalinist dogma of Engels's pronouncements in his *Dialectics of Nature*, like the biology of Lysenko, had caused his break with communism.

What we call "biochemistry" today did not receive its permanent definition until the difference between colloids and crystalloids was eliminated. But before being abolished this difference changed in meaning. In Ostwald's time, it could still concern chemists, even be the field of exploration for an eventual "new chemistry," beyond the science of molecular structures. Engels's thesis on permanent exchange as a condition of the existence of organic substances would then have heralded a chemistry centered on the distinction between the stable molecules of human laboratories and the "dynamic" molecules of living things. Not only did this physical chemistry never see the light of day,[7] but as early as the second decade of the twentieth century, it no longer interested physical chemists, who had become, as we shall see, atomists like their organic colleagues. Thus "dynamic biochemistry" would disappear unregretted, without the least repercussion in the world of chemists. The techniques of the chemistry of analysis and synthesis were certainly useful and necessary for the study of living things, but no respectable chemist any longer expected his science to contribute more than it already had: the notion of molecular structure and the study of the bonds that stabilize it. So the building of biochemistry finally established the connection between the chemical and the living, but the new discipline gave chemistry no other role than that of instrument.

A Physics for Chemistry?

In the last decades of the nineteenth century, physical chemistry was a plausible candidate to produce a new identity for chemistry. Was there to be a repetition of the situation created by Berthollet, when the authority of the Newtonian physical model seemed to put the identity of chemical bodies into question? Not quite, because the theory that served as reference model had changed. It no longer focused on motions and interacting forces; now the emphasis was put on a concept promising the unification of physics and chemistry, for at its foundation was the notion of "conservation of energy."

Chemistry, Energy, and Forces

Ever since the notions of energy and conservation began to share a common thread of history, chemistry participated in it fully. (In the beginning one spoke of "force" rather than "energy," as the respective meanings of the words were only slowly fixed; see Elkana, 1974.) Julius Robert Mayer (1814–1878), one of the inventors of "conservation of energy," even used as support for his law the difference in oxygen consumption in hot and temperate countries, which he said showed in the coloration of the blood of the inhabitants. Electrolysis and the battery were integral parts of the network of energetic transformations that linked all natural phenomena. They showed that electrical energy could be transformed into chemical energy and vice versa. Electrical energy could, moreover, be transformed into heat energy, which is itself consumed by chemical reactions, or produce light, which could itself determine chemical reactions, as in photography.

At the time electricity and heat were autonomous fields of study, distinct from chemistry, as are chemical, thermal, and electrical energy. But could one not envisage a new unity of the sciences, through which nature could be addressed in terms of energetic conversion? Could chemists construct an energetic theory of chemical transformation, just as mechanics, since its Galilean origin, had constructed an energetic theory of motion?

Marcellin Berthelot and Julius Thomsen (1826–1909)[8] of Copenhagen would devote themselves to the foundation of a "thermochemistry," a science relating in a systematic way, on the model of mechanics, the two most traditional actors in chemistry, the chemical reaction and heat.

Contrary to Berthollet's static chemistry, thermochemistry did not involve a hypothesis on the nature of the force that created the chemical bond. The old force, which was supposed to explain both the bond and the reactions, was part of a past that had to be overcome. Henceforth the model was the abstract formalism of mechanics, which stressed the ideas of work and energy. Just as the fall of a body is characterized by the work of mechanical forces, the decrease in potential energy and the creation of kinetic energy, a chemical reaction must be defined by the work of chemical forces and the decrease in potential of these forces. Work and decrease in potential were to be measured by the amount of heat released by the reaction. The state of chemical equilibrium thus became the state in which the potential of chemical forces had reached its minimum value, and it was defined by the fact that the reactions that led to it were those that involved the largest release of heat, which Berthelot stated in 1865 as the "principle of maximum work."

In a sense, as Ostwald noticed, thermochemistry was just a transposition of the old doctrine of elective affinities, which assumed the dominance of "the strongest affinity" (Ostwald, [1906] 1909, p. 213). This transposition corresponded, as in Bergman's time, to a discrimination among chemical reactions: since it was the release of heat that measured the work of the forces in a reaction, the natural chemical reaction was the one spontaneously giving off heat; endothermic reactions, which absorb heat, were considered *constrained* by an external action, by the chemist who adds the heat.

The principle of the conservation of energy was certainly respected in chemical reactions, but could it be used to predict the state of equilibrium—namely, the direction of the reactions produced from a given initial preparation and the concentration of the end products? This question was very important to specialists in synthesis, because it was the question of the *yield* of a reaction that was being asked. While chemical analysis favored powerful reagents (in Berthellot's terms, complete reactions), in synthetic chemistry the goal was to enlarge the variety of the reactions used to obtain the extremely specific transformations it needed. And a science that had the means to improve the yields of "incomplete" reactions would be welcome indeed.

As the jungle of known examples became denser, the difficulties of distinguishing between "purely chemical" (exothermic) processes and constrained (endothermic) processes were growing, particularly in the new area of high temperatures. Henri Sainte-Claire Deville discovered that water dissociated into hydrogen and oxygen when it was put into contact with a platinum sphere heated white-hot—which was odd, because the oxyhydric blowtorch, whose heat resulted from the combination of hydrogen and oxygen, was capable of making platinum *melt*. If the temperature of white-hot platinum was enough to "constrain" the endothermic dissociation of the water, the same should occur at the melting temperature of platinum, which was much higher. In fact, Saint-Claire Deville showed, the torch's temperature was lower than it should be if the combination took place alone. So the heat given off was not an adequate measure of the exothermic chemical reaction. It seemed rather to be a function of an "equilibrium" between combination and dissociation, two reactions that it became difficult to define as opposites, with one being purely chemical and the other constrained.

As early as 1867 two Norwegian chemists, Cato Guldberg (1836–1902) and Peter Waage (1833–1900), had proposed a law that abolished any distinction between exo- and endothermic reactions and created a new kind of analogy with physics. They put forward the idea of "active mass": the concentrations of reagents effectively present in the reactional milieu at a given moment and available for reactions. The "law of mass action" had the appeal of a mechanical law: like a Newtonian force, the chemical force of a reaction was defined as the product of the active masses, and equilibrium was reached when the forces of opposite reactions became equal. Unlike in mechanics, however, the active masses were variable, since they changed with the reaction and reached determined relative proportions when equilibrium was reached. Moreover, the relationship between forces and masses involved a specific "coefficient of activity" for each type of reaction. At equilibrium when the sum of the different chemical forces was cancelled out, the relationship between the active masses was given by the relationship between coefficients of activity, which were themselves dependent on temperature and pressure.

An even more profound difference between mechanics and Guldberg and Waage's law has to do with the effect of the force. While a true mechanical force determines an acceleration, the force introduced by Guldberg and Waage determined the rate of the corresponding reaction.[9]

At the time the reaction rate was a new phenomenological quantity. Only in the very slow reactions that occurred in organic chemistry could the variation in the concentrations of the reagents be followed over the course of time. Guldberg and Waage had made use of the results of the study of reaction times begun by Ferdinand Wilhelmy (1850) and continued by Péan de Saint-Gilles and Berthelot in 1860: at each instant the rate of a reaction was proportional to the concentrations of reagents still present in the reactional milieu; the rate therefore decreased as equilibrium was approached. The definition of the "chemical force" integrated the phenomenological study of rates with the model of mechanics.

The Kinetic Hypothesis

Guldberg's and Waage's hypothesis was a great success in chemistry as a phenomenological law. But it left open the question of its interpretation. Would it be possible for chemistry to adopt the kinetic hypothesis, produced some years earlier in physics?

In this hypothesis equilibrium was no longer defined as a state in which reactions stopped but as a state in which reaction rates were such that their effects *compensated for each other*. The kinetic interpretation of equilibrium had been suggested as early as 1857, by Rudolf Clausius, for the case of evaporation. Adopted by James Clerk Maxwell in 1865, it led to the idea of "Maxwell's demon," an image Maxwell used as a possible explanation of evolutions that took a system far from equilibrium.

The kinetic hypothesis sent the notion of "chemical force," however it might be measured, back to the level of analogies, since equilibrium was not the state where forces, and the rates they determined, both vanished. Equilibrium was nothing special: it was the state in which the "reactive collisions" between molecules, which determined a given reaction, were *on the average* as numerous as the collisions that determined the inverse reaction. The central concept of kinetics is the probabilistic notion of frequency. If an increase in temperature increases the rate of the approach to equilibrium, it is because it increases the frequency of all the reactive collisions. As for the questions of affinity, of the respective "forces" of the reagents, of the difference between endo- and exothermic reactions, of the composition of chemical equilibrium, they were outside the scope of kinetic theory and would be elucidated only through an eventual future science dealing with the reactive collision itself.[10] The kineticist could know only the number of molecules that participated in a collisional reactive event (which is called the "order" of reaction). This, by the way, was enough to occupy anyone. In effect, kinetic analysis would show that behind "a" chemical reaction a series of much more complex intermediate reactions was often hidden.[11]

Kinetics was a realistic hypothesis. It implied that chemists' molecules were clearly recognized as discrete entities, capable of motion, collisions, and individual behaviors—quite different from the convenient symbols used by synthetic chemists. And it opened quite a different perspective from the rival interpretation, produced at the same time, of chemical equilibrium and the law of mass action that defined it. This "thermodynamic" interpretation's central concept was indeed *the most abstract quantity* that nineteenth-century physics had defined, entropy, and it did rule out any intuitive understanding of chemical reactions.

Thermodynamic Equilibrium

The creation of the notion of entropy in physics was a response to a problem similar to the one that condemned Berthelot's and Thomsen's thermochemistry: the principle of the conservation of energy, always verified by physico-chemical transformations, could not be used to determine which transformations were possible and which were not. To the "first" principle, that of conservation, Clausius added the "second principle," which involved a new function, entropy. Spontaneous transformations that would conserve energy but diminish entropy were excluded. The state of thermodynamic equilibrium was defined by the fact that any spontaneous transformation that would affect it would act contrary to the second principle of thermodynamics, so none was possible.

In 1884, a twenty-three-year-old chemist, Pierre Duhem, submitted a doctoral thesis in Paris that applied Clausius's principles of thermodynamics to chemistry. This thesis was rejected by the members of the jury, who were bewildered by it (see Brouzeng, 1987). Not only had this bold young man allowed himself to criticize the great Berthelot's glorious principle of maximum work, but he had even proposed a mathematical representation of chemical reactions, which assumed that they neither released nor produced heat!

None of these chemists knew, apparently, that Duhem was just inviting them to reproduce in chemistry the idealization, quite as scandalous, that Carnot and Clausius had used to subject the steam engine to the mathematics of measurement. While the real steam engine functioned by heating and cooling, the function of the ideal steam engine necessitated the fiction that two bodies of different temperatures never be put in direct contact! The measurement of a steam engine's cycle was then given by a transformation that *conserved entropy:* the system was supposed never to leave the state of equilibrium. It underwent a displacement from one state of equilibrium to another infinitely close state of equilibrium, a displacement entirely determined by an infinitely progressive variation of the control parameters.[12] It was the same for the chemical reactions, which Duhem characterized as fictitious transformations, reversible and entirely directed from the exterior from one state of chemical equilibrium to another.

Duhem's thesis never having seen the light of day, it was van't Hoff who would be associated with the law of *displacement of equilibrium*.[13] In the same year, 1884, Le Chatelier proposed, without demonstrating it, what would become a "principle," the definition of the way in which a chemical system at equilibrium reacted to perturbations from the exterior. In 1893 Duhem, exiled to Bordeaux because of his stand on Berthelot's theory, published in Ghent (French editors judged it heretical) his *Introduction to Chemical Mechanics*. Attacking Berthelot's thermochemistry with implacable cruelty, Duhem demonstrated that all the "principles" suggested until that time were direct consequences of just one pronouncement: "The thermodynamic potential of a system is at a minimum when the system is in stable equilibrium." In 1897 an imperturbable Berthelot published his *Thermochemistry*.

Chemical thermodynamics therefore had for its central principle a no longer energetic but thermodynamic potential. The definition of chemical equilibrium as a minimum of the potential made it a direct consequence of the second principle:[14] all spontaneous evolution that would move the system away from the minimum value, defining equilibrium, of the potential would contravene the second principle. Van't Hoff's law of displacement could be deduced from this definition, as well as Guldberg's and Waage's law of mass action and the articulation between the different parameters—chemical composition, temperature, and pressure—that determined the state of equilibrium.

The two approaches, kinetic and thermodynamic, both defined physical chemistry as an autonomous science in relation to mechanist physics, which had neither reactional event nor second principle. In both cases the relationship to physics was not reductionist: the kinetic exploration of reaction rates, like the application of the two principles of thermodynamics (or of energetics, in Ostwald's terms) to more complex cases in physical chemistry should, on the contrary, have made chemistry a field much richer than physics alone, a field in which physical concepts would be generalized.

Each of these approaches, however, seemed to lead toward a different future. Kinetics, with its reactive collision, suggested a connection to physics; this connection implied an acceptance of the reality of atoms and molecules and attributed to them, beyond observable phenomena, the responsibility for the properties of chemical bodies as well as for the modalities of chemical transformation. On the other hand, chemical thermodynamics strengthened the positivist dimension of chemistry and distanced it from any intuitive representation of chemical phenomena and their causes to make them an abstract function of the manipulable parameters. The hesitation between these two perspectives did not belong to chemistry alone but was equally present in the physics of the time.[15] That is why we can say that at the turn of the century physics and chemistry were no longer strangers; both sciences were at the crossroads, confronted with the problem of their future, of their identity.

From the Chemistry of Elements to the Physics of Nuclei

At the end of the nineteenth century a new kind of phenomenon made its appearance in physicists' laboratories: rays. Cathode rays, studied by the English physicist, William Crookes (1832–1919), seemed to consist of negatively charged projectiles; X rays, identified by the German physicist Wilhelm Roentgen (1845–1923), were electrically neutral and very penetrating; N rays, spotted by the physicist René Blondlot of Nancy, did not withstand close scrutiny. But it was "uranic" rays, emitted by a salt of uranium and discovered in 1896 by Henri Becquerel (1852–1908), that would transform the relationship between physics and chemistry irreversibly. To whom did radioactivity belong? One might think this a silly question, a vulgar quarrel over appropriation that should interest no scientist worthy of the name. But radioactivity was at the origin of two historic series of events, at the end of which the interpretation of the properties of the chemical element would be considered to belong "naturally" to physics, the science of principles.

The Two Definitions of Radioactivity

The name Curie is as indissolubly linked to radioactivity as those of Ampère and Ohm, for example, are to electricity. A scientist rarely has the privilege of seeing his or her name attached during his or her lifetime not just to a discovery, a theory or an effect but to a standard of measurement, to the norm that allows researchers and technicians of all countries to have a common language. The event was even rarer because Marie Curie (1867–1934) herself set the definition of the "curie"—the activity of a gram of radium—in 1910 and prepared the sample of radium chloride, carefully purified, from which other secondary standards would be produced, allowing laboratories in other countries to measure the radioactivity of their products. The event was even more exceptional in that, at the time, researchers in radioactivity had another possible definition available, which would prevail in 1962: they substituted for the gram of radium any quantity of radioactive material undergoing 3.7×10^{10} disintegrations per second. The first definition arose from chemical practice—the purification of a compound and the characterization of an element. The second's object was an event inaccessible to direct observation, i.e., disintegration, and the subject of the event, a "matter" undifferentiated except by its radioactive properties, i.e., the physicists' matter and not the chemists'. This contrast is illustrated by two famous scenes.

The first took place one night in 1898 in Paris. A few months earlier Marie Curie had shown that the intensity of the Becquerel radiation depended only on the quantity of uranium or thorium, and that it was invariant in relationship to temperature, dissolution, the composition of the uranium salt, in brief to everything that chemists could vary. The radiation therefore seemed to refer to the uranium as an *element*. Marie Curie gave the name radioactivity to this new elementary property. She had also noticed that certain minerals, pitchblende and chalcolite, were four times more radioactive than they should be if the uranium explained their radioactivity. Then Pierre Curie gave up his own research in physics to collaborate in the painful work that Marie was undertaking. Isolating a new element from pitchblende meant repeating interminably the same type of operation: cut off twenty kilos of the mineral, grind it, bring it to a boil, dissolve, filter, precipitate, crystallize . . . But to crown such efforts by "seeing" the finally purified element is the dream of every chemist. That is why on that night in 1898, in the hangar they used as a laboratory, Marie and Pierre Curie contemplated with ecstasy the slightly luminous silhouettes, which looked as if they were suspended in the night, of the flasks containing a minute quantity of the radium salt finally extracted from tons of pitchblende.

The second scene took place in Montreal in 1902. Ernest Rutherford (1871–1937), a physicist from New Zealand, and the English chemist Frederick Soddy (1877–1956) had studied a curious property of thorium, an element discovered by Berzelius in 1838 and identified as radioactive by Marie Curie and Gerhard Carl Schmidt in 1898. Rutherford showed that another product could be extracted from thorium, which he called thorium X, and which contained all the radioactivity. But the peculiar thing was that the original thorium became progressively radioactive again! And again one could extract radioactive thorium X, and so on. The only possible interpretation was that the thorium was transforming itself into thorium X, as witnessed by the radioactivity. Thorium transformed itself slowly, so it could appear nonradioactive for a little while. Thorium X continued to transform itself much more rapidly into other products, so it remained radioactive. At this instant, Soddy remembered, something greater than joy overcame him: the exaltation of being, among all the chemists of all times, the one who realized the old dream of alchemy, transmutation—the transformation of one element into another. Rutherford scolded that if they talked about transmutation people would treat them like alchemists! Soddy's exuberance could not be quelled; he marched around the laboratory singing, "Onward, Christian Soldiers." Two scenes of joy: one serene, crowning a hard and repetitive labor; the other intense, celebrating the moment when the pieces of the puzzle fell into place. In the background, the two different definitions of the curie, and the redefinition of Mendeleev's chemical element, which would lead to its appropriation by physicists.

The Radioactive Elements

Marie and Pierre Curie's research program belonged to chemistry. Of course the electrometer previously invented by Pierre Curie and used by the couple to measure the intensity of the radiation belonged to the most up-to-date physics, but it was only an instrument: it could quantify the phenomenon by a property, the fact that uranic rays, like X rays, made the air an electrical conductor. The essential thing was the purification and the promise of a new demographic explosion of elements, similar to the one started by the use of the electric cell. Marie and Pierre Curie first isolated one element and called it polonium, then a new substance they called radium. Because it existed in sufficient quantity that the protocols of analytic chemistry were applicable to it, radium would be the first to receive its Mendeleevian identity card. On March 28, 1902, Marie Curie could write, "Ra = 225.93." The same year Rutherford and Soddy arrived at the conclusion that knocked off the interpretation of Mendeleev's table: radioactivity was not an elementary property, it was the index of the transformation of one element into another.

The origin of this revolutionary result went back to Pierre and Marie Curie's observation of what they called "induced radioactivity": a salt of radium or polonium in contact with a metallic leaf made it radioactive. Very quickly the German chemist Friedrich Ernst Dorn, for radium, and Rutherford, for thorium, showed that the radioactive source emitted a gas, a radioactive "emanation," which was deposited on the surface of bodies close to it. At the end of 1899 Rutherford measured the time during which the emanation remained radioactive when left to itself. At the end of 54 seconds, the activity had diminished by half; at the end of double that time, it was reduced to a quarter; at the end of triple the time, it was reduced to an eighth: which meant its rate of decay was exponential. The emanation of thorium was the first radioactive body to which a mean lifetime of 54 seconds was attributed. And in 1902 Rutherford and Soddy showed that the emanation from thorium was not produced directly from the thorium, but from a product of the disintegration of thorium, thorium X, and they had constructed the disintegration hypothesis.

Then, for all those who accepted this hypothesis, the problem of the identification of radioactive bodies took a new turn. If there were successive disintegrations, each disintegration product being characterized by a particular half-life, the purification of radioactive bodies could never be completed: radioactivity and purity were, in fact, mutually exclusive. So it was known as early as 1910, when the curie was defined by reference to the activity of a source of pure, radioactive radium, that this definition taken from the chemistry of elements was only approximate, relative to a peculiarity of radium. It happened that radium, a product of the disintegration of uranium, provides an ideal radioactive source, both intense and stable. In effect, its mean lifetime, 1,622 years, is neither "too long"— radium's activity is not too low—nor "too short"—radium's quantity is conserved compared to the time of chemists.

What, then, was the task of the radiochemist? It can be compared to finding a solution for a kinetic puzzle. Let us consider a salt of thorium.

Its radioactivity varies continuously during the years following its fabrication. Thanks to Rutherford and Soddy, we know what that means. No more than one can obtain a pure radioactive product can one, in general, measure the exponential decrease of a radioactive product directly. In effect, most often one measures the result of several processes that succeed each other for each particle, but coexist in the same sample. From then on the global radioactivity measured must be interpreted *in kinetic terms*. It depends on a series of distinct actors, of which each one is produced at a rate that depends on the entire preceding series and disappears at its own rate, producing a new actor. That is why, before reaching a global equilibrium, the reactional milieu is characterized by variations in the activity of radioactive emission as a whole, and of relative values of the emissions α, β and γ, which are the ingredients of the radioactive emission. These are the variations that the radiochemist must decipher. There are variations in activity that indicate that a product with a rather long half-life is present: if it can be isolated, it can be identified and its series of transformations down the line can be studied separately. Others disintegrate too fast to be isolated, but ingenious, specially developed devices allow them to be studied, i.e., their lifetime to be evaluated, and the type of radiation that they emit in disintegrating to be identified. Yet others are first postulated on the basis of indirect indices. At the end of the investigation, each atom in a radioactive series has been named, its atomic weight determined, its half-life evaluated, what it emits while disintegrating identified.

As far as thorium is concerned, the reconstructed history was finally written thus: in disintegrating thorium produces β particles (electrons) and mesothorium I, which produces β particles and mesothorium II, which produces β particles and radiothorium C, which produces α particles (helium nuclei) and the famous thorium X, which produces α particles and the emanation, which produces α particles and thorium B, which produces β particles and thorium C, which produces both β particles and thorium C′ and α particles and thorium C″, the two possible products finally producing the same stable product . . .[16]

The thorium series is one of the four radioactive series known, which all have almost the same look. Does that mean that Mendeleev's table was suddenly enriched with about forty new "elements"? Not at all. Except for a few newly filled squares (polonium, radium, radon), the table remained what it had been, but its meaning was transformed. This conclusion was worth a Nobel Prize for Soddy in 1922, Rutherford having received his in 1908. Each radioactive series wanders through the same region in the table with a similar gait: two squares back when there is an emission of an α particle (the nucleus loses two positive charges), one square forward when there is an emission of a β particle (when a nucleus loses a negative charge, it is as if it gained a positive charge).

One can thus follow each series on Mendeleev's table and realize that radiothorium inhabits the same square as thorium, that mesothorium I and thorium X inhabit radium's square, that thorium B is in lead's square, and thorium A and thorium C' are in polonium's square, that the thorium emanation is with radon . . . In 1913 Soddy named these multiple inhabitants of the same case isotopes, *iso-topos,* "which are in the same place." In each place in Mendeleev's table there was no longer just one element, but a certain number of distinct atoms, all having the same chemical properties, but distinguished by their atomic weights and the instability of their nuclei (their mean lifetime). All of a sudden Prout's old idea based on the nearly whole number values of atomic weights, which Duhem had made the symbol of the irrational temptation to make speculations violating the facts, was justified! Chlorine's atomic weight was 35.5? This element has indeed two isotopes, of atomic weights 35 and 37, the first being three times more frequent in nature than the second, so the average is 35.5.

Physicists in Quest of the Atom

Soddy identified isotopes. Rutherford accumulated "discoveries": in 1909 he had shown that one of the components of radioactive radiation, called α, consists of helium nuclei; soon he began bombarding nonradioactive substances with those particles and deduced a first model of the atom, a massive, positively charged nucleus surrounded by negative electrons in 1911. In this context the turn that Marie Curie's research took after the double shock of the Nobel Prize in 1903 (received jointly with her husband and Henri Becquerel) and the death of Pierre Curie in 1906 was remarkable. Marie Curie persisted in verifying the chemical identity of radium, which the aged and famous physicist, Lord Kelvin, had questioned.[17] In 1911 she received another Nobel Prize for having isolated pure radium, but this type of research already belonged to the past. Radioactivity had become part of the history of physics. Chemistry came into it only as a technique for identifying the isotopes produced by transmutation.[18]

It is always useless to try to rewrite history, but we must pause for speculation here. For the first time chemistry and physics were confronted at the same time with the same enigma, and in the end chemistry found itself defined as a technique in the service of questions asked by physicists. It is difficult not to see in Marie Curie's persistence in purifying radium while Rutherfod was throwing himself into the exploration of the atomic nucleus a tipping of the scales that was factual, symbolic and irreversible all at the same time, i.e., historic. Indeed the distribution of roles that then began did not ratify a preexisting difference but created a new image for physics. The history that went, with Rutherford, from the element to the atom proceeded from one fruitful conceptual shock to another: the atom was no longer *a-tomos,* since it disintegrated, Mendeleev's table no longer classified—it was plastered over with a cascade of transformations; elements actually lost their identity and regrouped themselves into an indeterminate number of distinct isotopes. Rutherford had dared to invent hypotheses that transformed the categories of chemistry into simple facts to be interpreted. Could this English boldness be related to the ancient but lively Newtonian tradition, which considered it obvious that chemical phenomena could be explained in terms of the physical interactions among bodies? In any case, physics was henceforth defined by a new challenge: to go beyond observable phenomena toward another reality that could interpret them.

From Atoms to the Atom

In 1925 the chemist Henry Le Chatelier wrote, in *Science and Industry,* "What is left of relativity, isotopes and quanta when they are stripped of the fripperies they have been dressed in? The same thing that is left of Perron chocolate when one takes down the posters that plaster the walls of Paris. It is chocolate like the others, which can be eaten without inconvenience; they are hypotheses like the others, which can be taken as a guide to research, but they are not discoveries" (Le Chatelier, 1925, p. 195). The chemist was on the defensive, but significantly he did not mention atoms among the fashionable hypotheses. Twelve years earlier, even the most famous adversary of the atomic hypothesis, Wilhelm Ostwald, had laid down his arms: atoms were not a simple model, a fiction of passing usefulness. He drew this conclusion from reading *The Atoms* by the French physicist Jean Perrin (1870–1942).

Atoms Exist!

The Atoms was meant precisely to force this conclusion—that is, to end the long hesitation caused by the atomist interpretation, a hesitation that took a new twist with the proliferation of kinetic models.

If two volumes of hydrogen gas (H_2) react with one volume of oxygen gas (O_2) to produce two volumes of water vapor (H_2O), it is because there are the same number of oxygen, water and hydrogen molecules in equal volumes of gas (at equal temperature and pressure). But what is that number? How many hydrogen molecules are there in 22.4 liters of gas weighing 2 grams? There are N molecules, but we do not know N, we know only the volume or weight *relationships*. That is the point where the chemists stopped. But the fact that there is the same number of gaseous molecules in the same volume at equal temperature and pressure can be interpreted with a kinetic model: gas molecules are fast, hurtling bodies whose size is ridiculously small compared with the gas container; their *average* speed is measured by the temperature, and their bumping into the container walls explains the pressure. In the "perfect gas," molecules move freely except when they collide with each other. If it could be taken seriously, this model could furthermore be used to interpret the empirical difference between a real gas and the perfect gas: the deviations would be explained by the fact that the molecules are not totally free from each other, and that their size must be taken into account, thus deviations would become quite interesting instead of just data to be recorded.

In fact, at the turn of the century kinetic models were no longer a simple way of interpreting already known experimental relationships. They were associated with cutting-edge physics, with the exploration of a world of discrete entities beyond the continuous, observable phenomena. As we saw, Rutherford's radioactivity required kinetics, but so did other fields. In 1887 van't Hoff had used a kinetic model to study osmotic pressure: he defined the pressure exerted by molecules dissolved in a liquid on their vessel as equal to that which would be exerted on the vessel by the same number of molecules in the gaseous state. Also in 1887, the Swedish chemist Svante Arrhenius (1859–1927) explained the electrical conductivity of saline solutions by assuming that the salt in solution was in fact dissociated into two types of independent particles, the "ions," one positively charged and the other negatively charged. The ions made it possible to unify the field of electrochemistry with the much older one of the chemistry of salts, acids, and bases. The force difference between acids or between bases was henceforth interpreted in terms of the "degree of dissociation" of the ions of these acids in solution, what we now measure by the pH. Furthermore, this degree of dissociation answered Guldberg and Waage's law of equilibrium. Cathode rays had also been associated with electrically charged particles, the electrons, whose relationship between charge and mass was identified by J. J. Thomson.

Finally, Brownian motion, the incessant and irregular movement of a light particle suspended in a fluid, offered evidence to the adherents of the atomic theory that this fluid, apparently at rest, did in fact consist of molecules in incessant motion (as chemical equilibrium was, according to the kinetic hypothesis, the result of incessant chemical reactions). Albert Einstein (1905) and Marian Smoluchovski (1906) developed the theory of this motion. They assumed, as Perrin explained it, that when the observation time decreases, the average velocity of the Brownian particles varies wildly in magnitude and direction without tending toward a limit: Brownian motion should be thought of as a *perfectly irregular* movement responding to a probabilistic law. In other words, this motion does not reflect mechanics and its trajectories but kinetics and its events.

Perrin made his debut with this new physics, notably with research on ionized gases. But after the work of Einstein and Smoluchovski, he devoted his research to an apparently humbler object, the study of droplets in suspension in colloidal emulsions. He pursued a single goal: to fix, beyond any possible doubt, the value of Avogadro's number, that famous N that appeared in all the kinetic models but that no one could figure out. In fact, it always appeared to be associated with another unknown quantity, the weight of the atom or the elementary charge of the electron, for example, or in other words, a quantity that defined the hypothetical elementary entity assumed by the model.

In the French context, the goal Perrin chose did not cover scientific questions alone. It was a question not only of silencing the French mandarins, who forbade all references to atoms, ions, or electrons in their territory, but also of demonstrating that science could perfectly well go beyond observable phenomena. It was a question of breaking a connection, promoted at the time by antirepublican Catholic thinkers (see Nye, 1972), between the epistemological definition of a science reduced to the sheer formulation of laws describing observed regularities without ever explaining them and the notion of the "bankruptcy of science," as a source of meaning and a search for truth.

There was only one possibility for finding a value for N: to associate two distinct kinetic models of the same phenomenon. In Perrin's time certain values for N had already been proposed. By crossing the kinetic model of the viscosity of a gas (which used the free average path length of gaseous molecules, the distance they traveled on average between two collisions) and the corrective hypothesis proposed by Johannes Diderik van der Waals (1837–1923) to the law of perfect gases, Joseph Loschmidt (1821–1895) and Rudolf Clausius (1822–1888) calculated this value for a series of gases. But to convince the antiatomists, Perrin wanted to find an experimental procedure that was above all suspicion. He found it with the emulsions, by crossing the theory of Brownian motion and van't Hoff's osmotic model: the nature of the liquid in which the granules of colloids were suspended, the size of the granules, and the temperature could all be varied systematically. If the value of N remained the same throughout these variations and agreed with already known ones, no one would be able to contest the possibility of "counting" the atoms.

So Perrin centered the question of the existence of atoms around the quest for N, the Holy Grail of the "atomist science." A reader of his book learned first to desire: if only we knew N, we would know . . . Then came the first evaluation, that of Loschmidt and Clausius, and immediately a quantitative view of the world of atoms was sketched out: "Each one of the molecules in the air we breathe moves with the speed of a bullet, travels in a straight line between two collisions about one ten-thousandth of a millimeter, is knocked off its course five billion times per second and could, in stopping, raise a dust particle visible in a microscope a distance equal to its own size. There are 30 billion billion of them in a cubic centimeter of air under normal conditions. It would be necessary to put three million of them in line to make one millimeter. It would take twenty billion of them to make a billionth of a milligram" (Perrin, 1913, p. 124). This was followed by other kinetic scenarios in serried ranks and other evaluations of N, in the first rank of which were the very precise measurements Perrin had made himself. At the end of his book, Perrin could line up thirteen similar values, obtained by thirteen independent methods.

Perrin's *The Atoms* is a rare example of a work of popularization whose publication affected the history of the sciences. None of the works presented was unknown to specialists, but for the first time they were assembled in one place and reinforced by their unanimity the conviction that a conclusion had been reached, ending a century of debate and affirming the identity and the reality of the chemical and physical agents used by kinetic models. Thirteen times the risk was taken and thirteen times the bet was won. No one could explain why N got the same value thirteen times without recognizing the actual existence of atoms (molecules, electrons, and ions). The *actual* existence of atoms took its place henceforth, next to the principle of the conservation of energy (which also evolved from putting disparate phenomena together), among the "discoveries" that physicists could confidently believe owed nothing to human prejudices and opinions and everything to nature—put another way, they were "true."

When Perrin defined the "atomist science" as the intuitive understanding that "explains the complicated visible by the simple invisible," he was only partially defining his own project. On the one hand, as he would show in the following pages, his project's direct consequence was to turn the simple visible into a complicated invisible. The regular, observable trajectories of mechanics were no longer representative of the behavior of matter. Instead, mathematically pathological trajectories (what we now call fractals) had to be envisioned. Just as the "trajectory" of the Brownian particle is infinitely fragmented, the properties that, on our scale, appear regular and continuous—and that, as such, have been made the object of the laws of physics—are in fact irregularity itself. The atomist does not reduce the complicated to the simple without at the same time revealing the complication of the simple, the hidden swarm of events and actors beyond the regular laws, which are therefore only human, artificial creations. On the other hand, the event in itself was not simple. From then on atoms existed for physicists as for chemists, but what about the individual atom, which could be expected to account for the formation of chemical molecules and for the reaction? On this subject kinetic models were mute. They involved events defined by their

frequencies and by all the circumstances that influenced those frequencies, but they delivered no theory of the events themselves.

Bohr's Model

In that same year of 1912, when Perrin finished his book, a young Danish
physicist, Niels Bohr (1885–1962), who was working in Manchester with
Rutherford, undertook the work that would lead to the first atomic
model capable of reconciling the general laws of mechanics with the
specificity of the chemical element.

The construction of Bohr's atom belongs to the history of physics.
Physicists often relate the story through a narrative thread that follows
the controversies on the nature of light and ends in Einstein's hypothesis
on luminous quanta and de Broglie and Schrödinger's wave mechanics.
This narrative hides the role played if not by chemists at least by the
knowledge that chemists had produced on atoms and molecules; in
particular, it ignores the body of knowledge that led chemistry,
strangely, to the arithmetic of discrete quantities and not to the calculus
of continuous functions.

It is clear that the young Bohr's object was not to submit chemistry to
mechanics. Like Perrin, but following another route, he intended to
bypass *both* mechanics and phenomenological chemistry. As it happened, he expanded Rutherford's model of the atom, trying to connect
the chemical specificity of the elements grouped in Mendeleev's table
with the description of the behavior of electrons, which, in that model,
spin around the nucleus. To a physicist, the main difficulty with Rutherford's model was that the energy of the electron, if it were spinning
around a positive nucleus, should be dissipated progressively, like that
of all charged bodies moving in a field: the atom should not be a stable
structure. Bohr was not too anxious, however: he was not trying to
reduce the atom to classical electrodynamics, but to define precisely the
transformation the chemical atom forced upon classical physics. He
knew from the start that the way atoms absorb or emit light energy was
not classic: Max Planck (1858–1947) had shown that the radiation from a
black body implied the discrete character of energy exchanges.

In 1913 Bohr returned to Copenhagen and moved all the pieces in the
puzzle around. The periodic character of Mendeleev's table could be
explained by the limited number of electrons occupying the same orbit:

when an orbit was filled, one moved down a line in the table. As for Planck's quantum discontinuity, it agreed marvelously with spectral emission rays and luminous absorption rays. Since the work of William Wollaston (1776–1828) and Joseph von Fraunhofer (1787–1826), these rays had been known to be an additional ingredient in the identity of the chemical elements: each element had its own distinct and terribly complicated spectrum. Johann Balmer (1825–1898) of Switzerland and Johannes Rydberg (1854–1919) of Sweden discovered a mysterious code that translated this complication into an arithmetic formula that generated a series of numbers, which turned out to be the entire spectrum of some elements, frequency after frequency. When Bohr discovered the formula for the hydrogen atom proposed by Balmer, the pieces of the puzzle came together: each emission or absorption ray was determined by the change in orbit of an electron, and the different orbits had discrete energetic values whose unit was Planck's quantum. Bohr's atom was born: the positively charged nucleus surrounded by electrons arranged in successive orbits—each with a distinct quantum number—and capable of "jumping" from one orbit to another by absorbing or emitting a quantum of light, this quantum representing the difference in energy between the two orbits.

Bohr's atom was nurtured by chemistry and breached the laws of physics, which made it possible to understand neither the stability of the electrons' motions in their orbits, nor the discrete character of the energies in these orbits, nor the instantaneous jump from one orbit to another. It was a question of postulates, which introduced the arithmetical regularities characteristic of the chemistry of elements into physics.

The Electronic Theory of Valency

Very quickly chemists started appropriating Bohr's model, because it yielded an astonishingly intuitive view of their results. The atomic number recording the position of each element in Mendeleev's table was nothing other than the number of electrons in the atom of this element. It was also equal to the number of positive charges in the nucleus. Each successive element in the table therefore had one electron more than its predecessor, and the periodic valence changes observable in the table reflected the successive filling of the orbits. The reactivity of the atom was determined by the outer shell, which was incompletely filled except in the case of the rare gases. Only atoms of rare gases could therefore be stable in the free state. The only stable combination was one that allowed the different atoms involved in the transaction to complete their outer shells. A new research field opened up for chemists, one that brought back in an unexpected way the traditional distinction between chemical reactions and physical interactions. The chemical body, henceforth the atom, was not indifferent to the configuration in which it found itself. The configuration had to satisfy a requirement that defined it: to complete its peripheral crown, to reach, thanks to chemical bonds, the point of "satiety," which would henceforth define the rare gases.

As early as 1916 the German chemist Walther Kossel (1888–1956) proposed an interpretation of ionization on this basis: each ion "completed" its peripheral crown, thus adopting a configuration similar to that of the nearest rare gas, the negative ions by the gain and the positive ions by the loss of one or more electrons. The same year, Gilbert Newton Lewis (1875–1946) of the United States enunciated the distinction between a polar bond (ionizable) and a non-polar (covalent) one: molecules joined by non-polar bonds are not dissociable into ions, there is a stable interpenetration of the peripheral electron rings of two atoms, so that each one could "saturate" its ring by sharing electrons. A bond meant that two electrons had been used in common by two atoms (electronic doubling) (Kohler, 1979; Vidal, 1989; Palmer, 1965).

Of course, Lewis's interpretation did not explain all Mendeleev's elements: Lewis submitted only thirty-two to his "octet rule," according to which a level was saturated when it contained eight electrons, and the question of the variability of valences remained unsettled. But in 1923, when the Faraday Society organized a "General Discussion on the Electronic Theory of Valence," its president, Robert Robinson, expressed the general feeling of the participants—a combination of confusion and elation: "At present it would seem as if we are in a transitory stage of knowledge comparable with that which obtained in the forties and fifties of last century, and the chemists of two or three generations hence will look back on the present confusion with the same feeling as we experience when regarding that time" (cited in Palmer, 1965, p. 141).

Indeed, things moved very quickly. The mathematical study of Bohr's orbitals by the German physicist Arnold Sommerfeld (1868–1951) introduced the possibility of elliptical orbits and the correlative necessity of characterizing each orbit by two quantum numbers as early as 1916. In

1925 the study of the fine structure of the spectra of alkaline metals led E. Stoner to introduce a third quantum number. The arithmetic of the successive levels kept increasing in complexity as in the same year spectral analysis of atoms in a strong magnetic field (the Paschen-Back effect) allowed Wolfgang Pauli (1900–1958) to propose a fourth quantum number and to present what would be called his "exclusion principle": two electrons in an atom cannot be characterized by an identical group of quantum numbers. With four quantum numbers, n, l, m, and s, all the elements in Mendeleev's table could at last be described in terms of orbits. The diversity of the electron configurations was the workshop in which increasingly precise links between reactivity and the atomic model were now being forged.

In particular, the idea of the "force" of a reagent (its capacity to displace another, which was the object of the affinity tables in the eighteenth century) and the great distinction between acids and bases were cleared up: it was the configuration of the electrons that determined them, and more especially the attraction between the electrons and the nucleus. The force of an acid, which Ostwald and Arrhenius had correlated with its degree of dissociation in the 1880s, could now be interpreted, and the periodicity in the level of reactivity of the elements in Mendeleev's table became comprehensible. The further to the right an element was located, the more "electronegative" it was (the more it attracted electrons). And the larger the difference in electronegativity between two elements united by a chemical bond, the more susceptible the compound was to ionization (to becoming dissociated into two charged ions).

To characterize the polarization of a chemical bond, chemists still use the scale of electronegativity built up experimentally by Linus Pauling of the United States and published in 1932 in *The Nature of the Chemical Bond*. And they constructed quite a few other tables, always experimental but now theoretically intelligible. With the "redox potentials" table, in particular, they had a real equivalent of the affinity tables. The redox potential is a measure of the difference in electric potential as measured in electrochemical cells in which oxidation (addition of electrons) and reduction (loss of electrons) reactions take place in separated compartments. It measures the gain or loss of energy corresponding to a change in the electron configuration of an atom—that is, to the loss or gain of one or several electrons. One can predict the direction of a reaction according to the difference in energy of two redox couples engaged in the reaction. Newton could now be transposed easily: metallic zinc put into a solution of copper salt (containing Cu^{++} ions) is dissolved (formation of Zn^{++} ions), while the copper is deposited (reduced), because the reduction potential of the copper ions is greater than the reduction potential of the zinc ions. Beyond the theoretical interpretation, the modern tables are distinct from the old ones in that they no longer determine a relative order among reactions that "measure" each other; instead, they provide a quantitative scale implying a common measure by a third party, the difference in electrical potential. So was Guyton de Morveau's dream fulfilled.

In about twelve years a new creature was born, an atom capable of bonding physicists and chemists, of teaching them both to understand facts that would henceforth be common to them in the same way. Not only did the progressive filling of the electron rings explain Mendeleev's table, which in turn guided the representation of the orbitals, but the atom that emerged from Bohr's model fit naturally into the language of kinetic theory, which, since Jean Perrin, had been common to physicists and chemists alike. In effect, in 1916 Einstein related the "quantum events" that constituted the electron jumps and Planck's law of black-body radiation, leading to a kinetic description of the probabilities of electronic transition. So Bohr's atom related Jean Perrin's atoms to Mendeleev's elements.

Science Deduced, Science Reduced

The connection made by Bohr's model between elemental chemistry, Perrin's kinetics, and atomic physics remains in place today, but its status has changed. Although in the beginning it played a constructive role and was recognized as an important part of the image of atoms, at present it belongs to the technical side of the work of physicists and chemists, little popularized and little commented upon. It is as if Fontenelle's judgment had come true. Physics is the clear and distinct science of principles. Chemistry is subordinated to it; it is based on approximations and therefore cannot, by definition, take part in the conception of principles. The only difference is that chemists are no longer the only victims of this hierarchical order: all the users of kinetic descriptions—all those for whom the probabilistic event is independent of observation—are in the same position. The kinetic theory is now connected to quantum mechanics through approximations (see Cartwright, 1983). The irregular world of Jean Perrin's atoms has disappeared in favor of a "quantum reality," which is unobservable but regular, because it conforms to a law.

A World without Events

As early as 1921 the physicist Max Born (1882–1970) wrote to Einstein that "quanta are really a terrible mess." The ability of Bohr's atom to interpret Mendeleev's table and explain valences, the dream of chemists, was a nightmare for physicists. Already two of Bohr's postulates (the stability of orbits and the quantum jump) seemed to be ad hoc hypotheses, but with the exploration of atoms with multiple orbitals the ad hoc hypotheses multiplied: it was as if physics, at the behest of chemistry, was forced to learn from the atoms how to understand them instead of deducing them from fundamental principles! Of course Bohr's "correspondence principle" served as a crutch, but it was insufficient to order the hypotheses.[19] They proliferated as experimental facts accumulated and diverged more and more from classical electrodynamics. By 1923 Bohr, Pauli, Heisenberg, Born, and many others were agreed: a radical change was necessary. Mechanics and classical electrodynamics, even when modified by Bohr's postulates, were stymied. A new theory was needed, one that would break the bonds that Bohr's model had retained.

The new quantum theory, proposed in 1927, is essentially our current theory of the atom. It has certainly created a new and fruitful coherence between facts and principles, but it has also destroyed any appearance of symmetry between physicists and chemists. The asymmetry is imprinted in the quantum formalism itself, because this formalism creates a dissymmetry between the description of the stationary atom (or rather of the idealized, stationary hydrogen atom) based on Schrödinger's equation and the description of events capable of affecting it. Such events cannot be deduced from Schrödinger's equation and seem to imply the intervention of a factor extrinsic to the quantum system, properly speaking. This is usually expressed through the association between the "reduction of Schrödinger's wave function" from which the probabilites of events can be deduced and the measuring apparatus that can record these events. It is as if it were henceforth forbidden to say that an atom emits a photon if there were no apparatus to record the impact of this photon.[20]

As in Ostwald's time, certain physicists have been speculating for a half-century on the disappearance of matter. Indeed, matter seems to owe its observable properties only to our observation. The possibility of such speculations shows that, when it's a question of "pondering principles," the testimony of chemists is null and void, unworthy of restricting the reflections of physicists. For chemists, whatever the method used to derive the probabilites of events in the Schrödinger equation, they owe nothing to observation, because if it *were* the case, no chemical reaction, no evolution described in kinetic terms—in sum nothing that interests the chemist—would have any existence independent of observation! A physicist may well imply that his stature permits it, while a chemist can only shrug his shoulders in silence over it. We are no longer in Venel's time.

Deduction or Coadaptation?

Here history tangles up several heterogeneous dimensions, which are difficult to contrast as "scientific" and "ideological" because ideology is not just an a posteriori commentary but confers its meaning and its focus on the history called scientific. Like Bohr's theory, the new quantum mechanics has strong bonds with the chemistry of elements and valences. Schrödinger's wave function allows us to represent in an exact way an ideal version of the hydrogen atom and, thanks to adequate approximations, it can help to construct the representation of the other chemical atoms and the interpretation of the variability of valences. As for the hydrogen molecule, Walter Heitler (1904–1982) and Fritz London (1900–1954) of Germany showed, as early as 1927, that it could be represented as a superposition of two wave functions. Since the two electrons belong equally to the two atoms, Pauli's exclusion principle applies to the molecule as a whole. This established an unexpected link between the theory of chemical bonding and the exclusion principle: the parallel spin configuration (fourth quantum number), which corresponds to a repulsion between atoms, is excluded. The creation of this link would symbolize the reduction of chemistry to quantum mechanics. London wrote in 1927: "If there were no electron spin, the Pauli principle would permit (to give Schrödinger's equation) only the anti-symmetrical solution, with repulsion between atoms, and the homopolar bond would not exist. From the point of view of the Pauli principle, factual homopolar chemistry seems to depend entirely on the existence of the electron spin."[21] Through Pauli's principle and the point of view that it authorizes, the physicist has therefore won a position of authority. Without the spin, no chemistry. We recognize the typical irony that announces a reduction operation in this dramatic pronouncement: the very possibility of the reduced science depends on an apparently secondary element of reductive science, just a question with the solution of an equation.

Not all physicists adopted a reductive style, certainly. In 1931 Heisenberg wrote: "The theory of valency, which was given by Heitler and London and by others, has the great advantage of leading exactly to the concept of valency that is used by the chemists. But it seems questionable to me whether that quantum theory would have found or have been able to derive the chemical results about valency, if they had not been known before" (cited in Palmer, 1965, p. 125). Heisenberg judged that the situation was rather that of coapprenticeship, but he still underestimated the contribution of chemists, because it was not only the facts on valency that had to be known. The facts of chemical thermodynamics on heat emitted or absorbed during reactions also had to be used to test orbital models. In fact, even in the simple case of the hydrogen molecule, Heitler's and London's model did not give the experimental value of the dissociation energy; it provided a clearly smaller value, as if the model overestimated the repulsion between the two electrons. So terms had to be added to the equation to represent the "ionic" case in which the two electrons belonged to the same atomic orbit. In the mid-thirties Linus Pauling named this mixture of different representations "resonance."

"Quantum chemistry of the elements" was therefore not deduced from quantum mechanics but rather reconstructed from disparate elements, some from quantum mechanics and others from data theorized by chemistry. But this "mixt" was defined in a hierarchical way for a long time. Thus, until the fifties, Heitler's and London's theory was considered more rigorous by many chemists, because it had a more direct link to quantum mechanics than the rival representation, that of the English physicist, John Edward Lennard-Jones. The latter, which starts from the hypothesis of molecular orbitals instead of from the superposition of atomic orbitals, is, however, the only theory applicable to complex molecules like benzene.

When it became clear to all chemists that the level of approximation required by the two theories was essentially the same and that the preference for Heitler's and London's theory was pointless, the controversy among chemists ended quietly. Specialists knew that henceforth the construction of a representation of orbitals would necessitate intuition and know-how above all. The art of coadapting experimental data with a pertinent version of quantum representation was needed.

We can therefore say that the quantum chemistry of elements, officially reduced to quantum mechanics, nevertheless kept an air of the past to the extent that the theories have not been "discovered" but have been negotiated in a relationship of coapprenticeship with the facts. But the situation is perceived differently. By their reference to quantum mechanics, the chemists' theoretical representations appear to be *approximations* with respect to an ideal knowledge, which is inaccessible because of the limits on our possibilities of calculation. The exemplary model of a practical science, resistant to the Newtonian dream, has given way to the image of a science struggling with objects that are obscure and difficult, certainly, but not very fundamental.

It is not without interest to recall that at the time when chemists gave up defining a molecule other than as a mixture of different idealized representations, Soviet specialists in structural chemistry were attacked by Stalinist ideologues for being corrupted by a "bourgeois, idealist, agnostic as to reality, Machian philosophy" (see Graham, 1974)! Didn't the structural chemists—satisfied with a theory that reduced the description of the chemical bond to a mathematical instrument, a fiction invented by the physicist or the chemist for his own convenience—give up determining what is really a molecule to content themselves with functional but artificial representations?[22]

At least the Stalinist ideologues were original in that they addressed themselves to chemists as if they were, like Kekulé and Butlerov, masters of their representations. In contrast to the debates from that period to the present in which the possibility is disputed of reducing the description of living things to the laws of physical chemistry, or the description of "mental states" to that of the brain, one point seems settled: chemistry has been *effectively reduced* to physics. The emergence of the chemical properties of the molecule from the physicists' quantum atom represents the first successful step in a conquest that, from level to level, would affirm the power of physical laws everywhere. The fact that the construction of the quantum atom, radioactivity, and kinetics had to be "negotiated" and not deduced directly from laws that flowed directly out of physics alone is treated as anecdote.

Correspondingly, after the first decades of the twentieth century, one no longer finds histories of chemistry written by chemists. There is no longer room to advance a thesis on the identity of chemistry, to draw lessons from it. Chemistry as such has no historians but professionals. As in the writings of Primo Levi, it is seen rather as a fabric of interesting situations, integrating the characteristic turn of mind of the chemist and the peculiarities of matter.

A Science without a Territory?

The image of a science is a product of its production style and scope, because the image enters into the ambitions and anticipations of those who choose to engage in the science. Those who choose chemistry for a career in the twentieth century know that their science falls under the shadow of Fontenelle's now prophetic judgment: where access has been acquired to a world ruled by intelligible laws beyond observable phenomena is the territory of physics; where the consequences of the intelligible laws of physics lose their clarity, where the obscure cuisine of approximations and negotiations with empirical facts begins, one has to do with chemistry. Even worse, they know that for many physicists the distinction between the two sciences is only a convention, a question of habit but also of differing interests. The status to which nineteenth-century chemistry aspired as a "positivist science," the demanding dynamic solidarity between experimental exploration and the construction of the object as reinvented by social and economic utility, has turned against chemistry. Following Fontenelle's judgment, chemistry may seem to be a kind of applied physics, whose focus is not the progress of knowledge but technico-industrial utility. Are not the new hypercomplex molecules constructed by contemporary chemists always presented to the public as promises of new technico-industrial processes, not as interesting in themselves?

Chemistry Everywhere and Nowhere

Certainly "great histories" will still be written whose heroes are chemical molecules—stories of enzymes, of desoxyribonucleic acid and the genetic code, for example. But these heroes will be described by specialists who, first of all, do not consider themselves chemists and, secondly, are interested in their molecules from the point of view of their biological functions much more than from that of the chemical properties they illustrate.

In a sense, the functioning of living beings has been reduced to chemistry, because the chemist's tasks—purifying molecules, reconstructing biological processes by mixing purified constituents—have accompanied the development of molecular biology at every step. In his story of the discovery of the double-helix structure of desoxyribonucleic acid (DNA), however, James Watson said that it was rather by accident that a chemist in the next office furnished him with an important piece of the puzzle: he should not rely on the usual textbook representation of the "bases" forming the chain whose structure he was attempting to understand. And the biochemist Erwin Chargaff, who in 1948–1949 had brought to light the strange regularity of proportion among the four bases (adenine, guanine, cytosine, and thymine) that characterizes all the DNA molecules, was able to gauge the incompetence in chemistry of James Watson and Francis Crick, who were researching the structure of this enigmatic molecule. "It was clear to me that I was faced with a novelty: enormous ambition and aggressiveness, coupled with an almost complete ignorance of, and a contempt for chemistry, that most real of exact sciences" (Chargaff, 1980, p. 102). But for Chargaff, besides the triumph of two ignoramuses in matters of chemistry, it was especially the "poverty" of the solution to the mystery of the regularity he had discovered that was traumatic: "What I did not want to acknowledge is that nature is blind and reads Braille" (Chargaff, 1980, p. 98). A technological kind of understanding had replaced chemistry as the key to biological processes. Between the chemical molecule and the biological macromolecule, there is no new general principle but only the prodigious diversity of the enzymatic "machinations" whose explanation and reason go back to biological evolution. Correspondingly, the chemist is a "service" provider from now on; he furnishes data and constraints, but he is no longer master of the situation.

Another factor contributing both to the rapid development of knowledge in chemistry and to the dismembering of its territory is instrumentation. The time when chemists weighed up the meaning and the relevance of the tools provided by the study of crystalline structure, specific heat, electrochemical data, or thermodynamic regularities is long gone. Today the chemist's laboratory is full of instruments that put physical theories to work, notably the instruments of analysis based on the interaction between matter and light.

In 1965 the first structure of a crystalline enzyme, lysozyme, was calculated from several tens of thousands of measurements of X-ray diffraction, the technique and method of calculation invented and used by physicists. Since then, information characterizing the individual constituents and individual events is more and more often substituted for the thermodynamic data, concentration, temperature, and pressure that defined the reactional environment as such. Thus, the technology of radiation (laser for monochromatic radiation, synchrotron for short and intense emissions, and so on) and the electronic analysis of signals is now multiplying access to stable molecular structures and complex transitory products during a reactional event. The chemist can now "see" what his ancestors thought would be forever a matter for speculation. The possibility of reconstructing, on the scale of atoms, the relief of crystalline microsurfaces allows him to envisage the analysis of chemical surface reactions and, perhaps, that of the catalytic reactions which seem to relate the reactivity of molecules and the structures of surfaces. He can also follow the steps of a chemical reaction in quasi-"real time."

But is this chemistry? During the last few years, scientists have succeeded in "filming" a reaction catalyzed by an enzyme, phosphorylase. When Louise Johnson, professor of molecular biophysics, explained the work that was done for this film, she first described the combination of the technique of X rays produced by synchrotron and von Laue's method of crystalline diffraction, and then wrote: "But we had to wait a little longer for the first real movie. To study an enzyme-catalysed reaction in progress, it was necessary to enlist the help of chemists" (Johnson, 1991, p. 33). In fact it was chemists who invented the means of "blocking" the molecule substrate by a photolabile (decomposing in light) group in such a way that the reaction and the start of X-ray measurement could be synchronized. This was the modest contribution identified as "the help of chemists" in the accomplishment of a chemist's dream, to "see" a chemical reaction.

Correspondingly, many chemists today admit that they no longer "dream." What determines their research now is the availability in their laboratory of this or that piece of costly and sophisticated experimental apparatus, whose use determines their project.

New Questions

And what if everything has not been said? Since the beginning of the seventies, a new element has complicated the logic of the story that the history of chemistry induces. Could chemical phenomena, if not reoccupy a territory, a base for the development of a stable relationship with neighboring disciplines, at least become—or become again, because the question leads us back to the time when chemistry was called the "science of the mixt"—a "field"? Could chemistry be a field in the sense of a ground for original questions and interests? Such a perspective assumes that, in contrast with Fontenelle's judgment, chemistry is not to be identified with the "confusion" in which its principles find themselves but can assert, on the contrary, that the operation of reduction to these famous principles allows a problem to escape, the problem of "chemical activity."

In this case it can no longer be a question, as in Venel's time, of the difference between a "mixtive union" and an aggregation, because the reduction operation has already been through that. On the other hand, an unexpected event marked the last twenty years: the question of the "macroscopic level," defined by the double route of approach, thermodynamic and kinetic, was revealed as a field of research whose fruitfulness was unsuspected from any microscopic model. The difference between the individual chemical reaction and the behavior of a *population* of molecules subjected to many simultaneous reactions is the focus of a new physico-chemical approach to matter. How can collisional events that are produced by chance encounters in the disorganized milieu of billions and billions of molecules in motion produce a coherent group behavior? This question is illustrated by the "chemical clocks" whose periodic color changes reflect a collective, global variation in chemical composition.

Like all unexpected events, this one has a modest origin. As Ilya Prigogine, founder of the thermodynamics of far-from-equilibrium systems, likes to tell the story, it goes back to his master, Théophile De Donder (1872–1957). An autodidact, specialist in mathematical physics, correspondent of Einstein on questions of general relativity, De Donder was instructed to teach a class on thermodynamics for engineers in 1911 at the Free University of Brussels. He therefore set out to define the status of this science, of which he was completely ignorant, and arrived at the conclusion that the problem that distinguished it was irreversibility. Now what are the intrinsically irreversible phenomena in nature, if not chemical reactions? At a time when thermodynamics seemed to be a backwater, De Donder quietly reopened the folder and reread it "backwards."[23] Instead of taking the classic approach of studying the state of equilibrium in which the reaction rates cancel each other out and entropy production vanishes, he posed the question of the chemical reaction as an entropy producer. He defined the contribution of each reaction to the production of entropy as the product of its thermodynamic affinity and its rate.[24] The production of entropy allows us to integrate into the thermodynamic approach the rate of chemical reactions, that is, to incorporate the temporal dimension of chemical activity into the formalism. It does not bring anything new to the definition of chemical equilibrium—the state in which the rates and affinities are simultaneously zero—but it modifies the whole thermodynamic perspective: equilibrium is no longer a privileged state, but only the state to which irreversible processes lead.

How to "leave equilibrium," to describe an open system whose relationship to the environment prevents it from going to equilibrium? For Ilya Prigogine, De Donder's student, answering this question was the prerequisite for an eventual thermodynamic understanding of "systems," which, like living beings, can be understood neither in terms of equilibrium (in the thermodynamic sense), nor in terms of evolution toward equilibrium (such evolution for living beings is identified with death and the arrest of metabolic processes). In other words, the question was to find out if Stahl's distinction between processes of degradation and processes that produce order, which had not been able to be verified on the level of structures (crystalloids and colloids), could have a thermodynamic meaning.[25]

As early as 1945 Prigogine established the "theorem of the production of minimum entropy," valid in the *near-equilibrium* domain, where the linear coupling relationships between thermodynamic "forces" (for chemistry, affinity) and rates established in 1931 by Lars Onsager (1911–1976), a chemist of Norwegian origin, were valid. In this domain the production of entropy plays the classic thermodynamic role of a potential function: its minimum, given the constraint that maintains the distance from equilibrium, guarantees the stability of the corresponding stationary state. It would take more than twenty years for the situation that prevails far from equilibrium to be recognized and defined where the second principle no longer allows a potential to be defined, and where, consequently, the question of the effects of the interplay of the various coupled reactions in the system can no longer be avoided. A new type of approach had to be created, incorporating the notions of the threshold of instability, bifurcation, amplified fluctuation—in other words, inverting the categories that prevail where the second principle can be used to define a potential. Although a potential guarantees that any fluctuation perturbing the state defined by its extremum will regress, far from equilibrium, it is necessary, on the contrary, to identify the kinds of systems that, at a given threshold, at a determined distance from equilibrium, become *unstable* with respect to a fluctuation.

Dissipative Structures

In 1969 Prigogine finally announced that, although they function far from equilibrium, certain systems—characterized by a nonlinear coupling between processes that produce entropy—are capable of "self-organization," of the spontaneous production of spatial differentiations and temporal rhythms. Prigogine called these coherent collective behaviors "dissipative structures": *structure,* because there was coherent spatio-temporal activity, and *dissipative,* because it occurred on the condition of maintaining the dissipative processes (that is, producing entropy).

In what way was this event unexpected? In the field of hydrodynamics, the existence of open and stable structures of the whirlwind type was already well known, and the discovery of the theoretical possibility of dissipative structures gave them, above all, their thermodynamic significance. Besides, the study of processes coupled in a nonlinear way was part, retrospectively, of a predictable development, because it was possible only by computer calculations and hence followed the development of the computer. The unexpected came from the role played here by the thermodynamic notion of an irreversible process and by the key position of chemistry in the creation of a link between the production of entropy and the production of coherence. In short, it led to the association of two sciences, thermodynamics and chemistry, which were cutting-edge sciences in the nineteenth century and thereafter considered passé, in the creation of a new evaluation of what matter can do and what its activity can produce.

In fact, in chemistry the possibility of dissipative structures was a surprise. The best-known case of a chemical reaction allowing the emergence of such a structure, the Belousov-Zhabotinsky reaction, bears the names of those who sought in vain to attract attention to its strange properties (variation in concentrations with a period of the order of a minute). Thereafter the "chemical clock" this reaction generates became a symbol, asserting that irreversibility produced a coherence that went back not to individual molecules that could be explained by quantum principles but to populations of molecules, to the coupling of the rates of chemical processes. This global coherence was manifested in the emergence of new descriptive parameters, on the macroscopic order of a minute or a centimeter, characterizing the molecular collective and not microscopic events and interactions. The coherence can also manifest itself in a "chaotic," unpredictable behavior. Spatio-temporal chaos and order are not opposites, but are both opposed to the incoherent disorder that prevails in the equilibrium state.

Far-from-equilibrium physical chemistry puts the accent on the global behavior of a population with local interactions. This question is as relevant in physics as in chemistry, and even in the study of some societies, as in the case of social insects, in which the behavior of individuals can be defined in a stable, separate way relative to the global behaviors that they engender. But chemistry is a privileged field of exploration for two reasons: first, because of the very great variety of cases, as much from the point of view of nonlinear couplings as from that of the temporal scales associated with each reaction; next, because the chemical processes create molecular structures, which can then exist independently of these processes (unlike hydrodynamic whirlwinds, for example). In other words, chemistry can produce stable structures that store a kind of memory of their formation conditions. It can thus act as a kind of relay between histories of different types: the physico-chemical history of the formation of material structures, the history that can be constructed on the basis of a property of these structures.[26] It is therefore the singular link between chemistry and a plurality of interwoven times, which was already the center of alchemical preoccupations, that makes chemistry a field for questions that the "reduction to principles" allows to escape.

Do we have here a resource for a "new image of chemistry," a science without a territory but one that can claim a field for experimentation on the borders? By contrast with the approach that "goes back to principles," chemistry could be the science that champions the interest of a new kind of "mixt," a mixt of processes, of levels of description, of regularities and circumstances. This is no longer a question for a historian; it can be answered only by those who make the history of chemistry.

EPILOGUE

This book is an attempt to meet the challenge of writing a panoramic, global history of chemistry, one that is not simply a list of facts and theories or an accumulation of anecdotes about individual chemists. Chemistry has been presented here as the real subject of a history that has developed through the constantly reiterated commitments of people, the knowledge they produce, and the meanings assigned to that knowledge.

We have seen pass successively across the stage a polymorphous, polycultural science, without definable borders in the world or in culture; then, a science annexing the chemical arts and occupying its own space, a niche in natural philosophy; next, a model of positive rationality and the basis of several prosperous industrial sectors; and, finally, a service science, subordinated to physics, and in the service of biology and industry, as well.

Chemistry and Its Image

Today chemists are sporadically undertaking to re-create an image for themselves in the eyes of the public. Colloquia, prizes, and publicity campaigns are promoting the value of chemistry. The injustice felt by chemists at being ignored, although their science is fruitful and useful, is shadowed by a certain perplexity, however: what image to present? The question is difficult, because the image that a science usually presents of itself is intimately associated with the idea of progress. What we have called the "dismembering" of chemistry's territory is a story in which progress plays the leading role. The dismembered territory was that of a science finally modern, a creator of general laws, standardized procedures, and standards of measurement. In order for the reign of laws to take hold, in order for the respectable profession to succeed the madman's science, chemistry broke its traditional alliance with heterogeneous materials, the products of processes that are resistant to forecast and dense with possibilities. And this "progress," which identified chemistry with the triumph of the power of describing, acting, and predicting, also made it vulnerable to those who could explain this power as well as to those who could use it. If there is a dominant image of chemistry today, it is of the exemplary illustration of the explanatory power of physical laws. The temptation to identify matter with what physical laws propose about it is as old as these laws, but it made Diderot and his colleagues, who found in chemistry the instruments for a cutting reply, laugh. The history of chemistry is the most impressive example of a coincidence of progress and reduction: the "vision of the physical world" no longer causes anyone to laugh today, whether this vision consists of speculations in which what we call "matter" is found to depend on the interpretation given to the quantum formalism, or it consists of "materialistic visions" of the brain that equate ultimate intelligibility with reduction to physical laws.

With a territory annexed by physics upstream, chemistry is, moreover, mobilized downstream to serve ends pursued by human industry. As a means of serving industry, chemical processes have a value determined by many factors—costs, profits, patents, markets—of which the chemist is not master. In 1965, the American Roger B. Woodward (1917–1979) could devote his speech at the Nobel Prize ceremony to the great saga of his synthesis of cephalosporine C. And he could still be presented as the outstanding second of Nature, the proud mother of creation. The syntheses of quinine, cholesterol, cortisone, strychnine, reserpine, chlorophyl, and more were then perceived as tours de force in which knowledge of reagents, astuteness, ingenuity, and imagination allowed the art of the synthesizing chemist to rival the synthetic mastery of life.

Today, on certain fronts, the chemist is equalling his "master," since he can construct molecules without equivalents in nature. The chemist can design biological enzymes like those "hollow molecules" that can capture and hold a molecule of a specific type—worth a Nobel Prize for Charles D. Pedersen, Donald J. Cram, and Jean-Marie Lehn in 1987. However, the art of the synthetic chemist no longer defines a stable identity: the synthesis of complex molecules is now most often conceived with help from computers with the cartography of all the paths to synthesis discovered thus far at its disposal.[1] Nowadays, the pharmaceutical industry synthesizes 25,000 variants of the same molecule. Since the beginning of synthetic chemistry, some ten million different molecules have been "invented," and this number grows by more than a thousand a day. The production of a new molecule is no more than background noise for other stories—and these stories do not relate first to chemistry as a science but to the interests and needs of industry. For one substance used by the pharmaceutical industry, nearly ten thousand have been tested and declared without intrinsic or commercial value. In other words, synthetic chemistry makes the "offer," but only the "bid" controls the meaning.

The theme of the "chance discovery" in pharmaceutical research illustrates the availability of chemistry to interests that it does not usually define, but it also symbolizes the frustration of this empirical kind of research: one cannot "yet" simply order chemists to synthesize molecules whose structure would be deduced, through the methods of biology, from their desired therapeutic function (see Pignarre, 1990). In the 1950s, a Rhone-Poulenc laboratory was synthesizing molecules with properties similar to those of histamine. One candidate had spectacular psychological effects: it was the first "neuroleptic," and its discovery, along with the identification of neurotransmitters and corresponding receptor sites, opened a new approach to the brain. But the work of the chemists who constructed these molecules, who isolated, identified, and determined the structure of the new neurotransmitting substances, was routine. The great story, the story of the "revolution" in neuroscience, was that of the biologists who identified *a posteriori* the relationship between molecular structure and biological activity; and it is to be expected that biologists will one day be the ones to determine *a priori* the therapeutic properties of molecules.

In the life sciences, as well, chemistry is generally treated as a "means," but there it is biological evolution that represents the ends. According to the classical theses on molecular biology popularized in France by Jac-

ques Monod in *Le Hasard et la Nécessité,* natural selection enslaves
chemical activity and contrives catalysis and regulation, which counteract the evolution toward equilibrium. Not new chemical laws but a
sophisticated chemical "technology" can be discovered from the study
of living beings. In any case, this is what is assumed when molecular
biology describes the way enzymes force reactions that otherwise would
take place at imperceptibly slow rates or regulate the rates of different
reactions. The secret of life being deciphered by molecular biologists
illustrates less the potentialities created by the chemical transformation
of matter than the quasi-technical "intelligence" that has subordinated
these transformations to a logic of survival and reproduction.[2]

Chemistry—present everywhere and nowhere, at work everywhere
but everywhere subordinated to questions, problems, interests, or techniques that do not belong to it—is moreover a victim of the contemporary avatars of the idea of "industrial progress." "Daughter of a
disreputable mother," alchemy, she had succeeded in constructing a
serious, moral, responsible image for herself by invoking her social
utility and her economic value. Now her industrial, agricultural, and
medical prowess, which seemed to guarantee the value of chemistry, has
turned against her and made her eminently vulnerable. The "professors'
chemistry" proudly displayed a coat of arms with a two-sided face: "pure
science" in the service of disinterested knowledge, and "applied science"
in the service of humanity. But today "purity" appears to be the preroga-
tive of physics, while chemistry has become the target in political and
social controversies over the values of industry and progress. Catastro-
phies like that of Bhopal (3,500 dead, hundreds of thousands irreversibly
handicapped), acid rain, chlorofluorocarbons destroying the ozone
layer, nitrate fertilizers and pesticides poisoning the groundwater, dangerous industrial wastes—all that is "chemical."

What can history bring to our understanding of this situation? In
describing the successive profiles that chemistry has presented over the
course of the centuries, we have tried to show that its present image as a
service science is not the mark of a progress toward a destiny but the
product of a history. Considering this present while turning toward the
future, we wish to use the lessons learned concerning the difference
between history and progress to suggest new representations of chemists
and of the knowledge they produce. A new image of chemistry is not the
affair of chemists alone: on it depends our image of matter.

Between Territory and Field

At the end of the last chapter we introduced a distinction between the notions of territory and of "field." To territory corresponds a power of definition, of delimiting, and this power inevitably creates the possibility of dismemberment. All it takes is for one power to be redefined by another power. A field, on the other hand, can be defined as a "ground for histories"; it is a theater for events and operations with sometimes necessary but never sufficient reasons. Field science cannot furnish the premises for a deductive approach, because purification does not prevail: the field scientist's work cannot be defined by the operations and manipulations he or she performs. Instruments can detect, locate, specify, and quantify but not define being in terms of operation. In the field, the scientist must learn over time what the locally relevant questions are.

The laboratory of the traditional chemist, the "madman's profession," was in itself a "field," for it did not have, as a general rule, the means to submit the products used to the purification that would assure the sorting out of the circumstances and accomplished, reproducible processes. In our discussion of corpuscles and atoms, we drew attention to the first act of the chemist whose ambition was to submit a chemical transformation to conclusive proof, the *reductio in pristinum statum*. At the end of the second chapter, we noted that the chemistry fostered by Lavoisier was circumscribed within a territory, the laboratory: the balance and the experimental closure that it implied established the power of the question of the multiplicity of circumstances. Or perhaps it only appeared so, because the skepticism of Lavoisier's contemporaries suffices to remind us that here, as elsewhere, the sharpest distinctions are the ones cutting through the most perplexing situations. In fact, as early as the nineteenth century, the limits of the conclusion became apparent, and chemistry, the model of a practical science, was forced to re-create new forms of multiplicity in the guise of elements, organic compounds, isomers, isotopes, catalysts.

Chemistry has not ceased seesawing between territory and field, between the predictive framework of theories that furnish *a priori* models and the exploration of the unexpected, of the multiplicity of cases. It continues to shuttle between the enclosed space of the laboratory entered only by purified beings, considered subservient to the power of the theories in whose names they labor, and the field, where the scientist does not in general have the power to separate *a priori* the essential from sheer circumstance, nor to expunge what the theories define as uncontrollable or parasitical. Today field-like diversity is proliferating more than ever in the very heart of the laboratory, even if it is marked out by protocols and standards of measurement. Some examples may illustrate this point.

Let us begin by exploring some physico-chemical far-from-equilibrium systems, since it is on their account that we have introduced the notion of field. This example will show how the chemical field can put in question the power of selective evolution, to which molecular biology submits it. Far-from-equilibrium chemical activity can produce by itself qualitatively differentiated, coherent regimes, with the properties of stability and instability. Accordingly, two contrasting styles of history may now define the relationship between matter and life: the "chemical" mutation history, which presents a succession of creative and improbable masterstrokes, or a process of intensification, of "modalization," of material that is already capable of many types of coherence.

In any case, some contemporary models of metabolic function are already enrolled in the latter perspective. These models pose the question of the distinction between features of a "purely chemical" metabolic functioning (the simple consequences of enzymatic regulations and catalysis, which are so many nonlinear couplings) and those that have been achieved by biological selection. We know, for example, that the breakdown of glucose takes place according to a time-periodic regime. In the traditional perspective of early molecular biology, this periodicity, since it exists, must be an effect of selection, and therefore it must have a precise functional meaning, even if we have not yet discovered it. According to the thermodynamic approach, the oscillation results from the coupling of the glycolytic reactions and may very well have no particular biological role (see Goldbeter, 1990).

How can the fact that an oscillating metabolic reaction does not necessarily have a functional significance define a new profile for chemistry? It is because the chemist, here, is no longer assigned to follow the contrivances of the power, in this case that of selective evolution: he also finds himself on the side of the matter whose activity is partially defined by this power but is not reduced to this definition. The necessity of learning, which is the very experience of field research, and the impossibility of making an *a priori* judgment have a positive meaning here, rather than marking the limits of our knowledge.

The theme of the "necessity of learning" is also, as we have seen in the preceding chapter, the end result of the "theory of resonance," the representation of a molecule by the construction of a quantum/phenomenological "mixt."[3] But apprenticeship does not just concern the negotiation of information and the constraints of distinct disciplines. More and more, what is important is the interest of the "case" itself, related to the definitional power of the rule. In fact, "cases" have proliferated in chemistry. All the great "laws" established by chemistry in the nineteenth century and retranslated into physical terms in the twentieth century now suffer exceptions. Today we are acquainted with "Berthollides," which have a nonstoichiometric composition, as Berthollet wanted. The typical notation for Berthollides identifies them by a little *deviation*, which separates them from a Daltonide: for example, $Zn_{1+x}O$, with x much smaller than 1. Berthollides are part of the vast category of *defects* in crystalline structure, compounds defined by their deviation from the corresponding ideal crystal.

Ideal, defect, impurity, deviation—these terms imply the usual hierarchy, the ideal being the prototype from which we can understand real things and their inevitable defects. But henceforth it is the "defects" that are interesting, for the specific properties they give the crystal. The crystal explodes into a mosaic of individual cases to which correspond various techniques for growing crystals meant to favor or avoid certain defects. The individual crystal is no longer an imperfect version of the ideal prototype but a reflection of the singular history of its growth. In other words, the multiplicity and often the industrial interest of the properties linked to "deviations" from the "rule" were substituted for the conventional differentiation between the "normal case" (an illustration of the rule) and "defects" (a nonhierarchical group of cases, each associated with the circumstances that favor it).

Each chemical element has also become a world in itself. One spectacular example is that of the rare earths. By 1939 all the elements in this category had been identified and their usual properties defined. Subjected to more and more highly perfected techniques for purification, carefully arranged in the framework of the periodic table, these elements nevertheless escaped the ideal of purification when chemists realized that their properties were ultra-sensitive to the presence of impurities. It is these impurities that make the rare earths interesting to industry: from the catalytic action in the cracking of petroleum[4] to the fabrication of television screens that reflect less light, by way of the fabrication of special metallurgical alloys, the rare earths have many advantages that make them industrial materials of primary importance. This relevance has, of course, transformed the profession of the rare-earth chemist. It is not only that some of them have been involved in industrial research. The chemistry of rare earths has also become a cutting-edge field in academic research, requiring expertise in quantum chemistry and computer simulation from its practitioners.

The computer itself has contributed in a powerful way to leveling out the difference between the general ideal and the singular case. Computer simulation is an innovation that affects all fields of science, but it is far from producing the same effects everywhere: specialists summarize this point in the salty slogan, "Garbage in, garbage out." In other words, sciences deprived of the means to construct a reliable model of the situation to be simulated will get from the simulation only what they put into it, an uncontrollable fiction with a scientific appearance. The use of this technique in chemistry sheds light on the unique epistemological status of this science. On the one hand, simulation creates a level of description intermediate between the microscopic level of the atom or the molecule and the macroscopic level, where most of the experimental facts are lodged. This "meso" level of a population of molecules/agents corresponds in many respects to the point of view that chemists contrasted to that of physicists as early as the eighteenth century: the contrast between the diversity of relationships and the regularity of general laws makes models of molecular dynamics a field of exploration, not a matter of the verification of the power of laws.

On the other hand, to the extent that it is a question of constructing a scenario using the principles that make sense of experimental facts, computer simulation illustrates the ambivalence of chemical phenomena between the rule of the laws of quantum mechanics and the circumstances that make a difference between explanation and deduction. "In the past," wrote Paul Caro, "there were theories on one side and experiment on the other, now between theory and experiment a third component, simulation, has insinuated itself . . . We witness the emergence of the idea that one cannot trust the experimental fact until it has been simulated . . . In chemistry we are proceeding toward a more and more abstract activity: the connection of instrumentation, calculation, image and manipulation of images" (Caro, 1991, pp. 211–212). Now if the activity in chemistry becomes more abstract, it also tends to escape subordination to the physical law. In effect, computer simulation does not respect the differentiation between the law and the ideal cases that obey it, on the one hand, and the approximations that ensure and verify its field of application, on the other. Simulation is a scenario in which legal constraints, particular circumstances, and their various interplays are all put on the same footing, translated into a single language specific to the situation. It therefore requires the art of negotiating a multiplicity of elements of knowledge, of manipulating them tactfully, and learning not to neglect a detail that could make a difference.

Thus, even in its avant-garde research, chemistry remains an art of the mixt, as Fontenelle would have said. But this coming together points up the contrast: the state of confusion in which the principles are found, far from being an obstacle, is henceforth the very heart of what matters in experimental and industrial practices as well as simulation. The difference is paramount. To confront the complexity of mixts, chemists today borrow tools from other disciplines—such as computer science, crystallography, or even biology, as when they use living organisms like machines to produce the molecules that interest them. In a word, if chemistry defines itself as a service science, it also profits from the service of others and networks with them in the service of mixts.

Between laws and circumstances, between territories and fields, chemistry always offers a space for great adventures. On March 23, 1989, two well-known electrochemists, Stanley Pons and Martin Fleischmann, announced that with a simple device—palladium electrodes immersed in heavy water—they had accomplished "cold fusion." Their story excited the scientific community and occupied columns in the newspapers. Certainly it ended badly, but in spite of its negative conclusions, the controversy over this claim instigated a rather significant argument (see, especially, Bockris, 1991). The partisans of cold fusion exposed and derided the confidence that physicists have in their laws. Cold fusion would have been the product of the chemist's art, of his patience and attention to circumstances. Suddenly the chemist became, as in Venel's time, the one who knew that the reproducibility of a phenomenon was not a natural property, which one could require of any scientifically respectable phenomenon, but rather the product of a long period of familiarization and apprenticeship. In the face of the certitudes of the "architects of matter," he again became the "dusty laborer" described by Diderot in

L'Interprétation de la nature, who "sooner or later brings back from underground, where he has been digging blindly, a morsel fatal to that architecture raised through brain power" (sec. 21).

In the last several years chemists have begun another adventure, which shows to what extent the "underground" illuminated by their theories still disgorges obscure bits that may offer the chance of windfalls. The hero of this adventure is good old carbon, the faithful companion that chemists believed they had known by heart since the nineteenth century but that had just revealed a new, secret existence to them. "Fullerenes"—from the name of the American architect Buckminster Fuller, who invented geodesic domes—are molecules in a stable arrangement of sixty carbon atoms in the shape of a spheroid like a soccer ball. This molecule, C_{60}, was discovered in 1985 during a series of experiments on carbon molecules in space. Since that time it has not stopped springing surprises. First of all, it has an extraordinary symmetry, with single and double bonds arranged into twenty hexagons and twelve pentagons. Moreover, it has an astonishing chemical versatility, and its physical properties, such as shock resistance and superconductivity, are no less remarkable. Finally, there is not just a single molecule of this type, but a whole class of molecules that seem to open up a new avenue for research. Since 1987 fullerenes have been all the rage and are featured in at least one article per week. The excitement caused in the world of chemists is comparable to that which welcomed radium. And, like radium, these molecules have the power to interest both theoreticians and practitioners, to reunite physicists and chemists in common enthusiasm. The fullerenes even have the marvelous property of introducing new customs into the scientific community. Next to the Nobel prizes and other honors for scientists, there is now an annual prize for the objects of their research. Elected "molecule of 1991," C_{60} seems to indicate that the individuality that is more and more difficult to find in the collective and anonymous work of research teams is more honored in the material they study.

No one can predict today what the future of the fullerenes will be. They are as likely to start a new field of research or a technological revolution, as the benzene hexagon once did, as to become a simple gadget, a "balloon" for the entertainment of scientists. But in any case these molecules are defining a new relationship between the investigator and the investigated, because they are not passive objects subjected to investigation but real partners, unique and individualized, of researchers.

Today chemistry complicates the image of progress, of which it was the best illustration in the nineteenth century. Certainly the most ambitious, promethean dreams come true every day. By penetrating the supramolecular world and playing with the interactions among molecules, the chemist has become a new kind of "architect of matter," a tricky architect for a tricky matter. Alloys with shape memory or photochromic glass are often called "intelligent materials," as if the chemist had breathed life into matter and accomplished the old dream of making the chemist's lab the triumphant rival of "the laboratory of nature." But there are other dreams now, other interests, which deeply transform the meaning of this demiurgic relationship between the chemist and nature. The fact that the great German firm Höchst has built a branch in Bombay to study the reasons for the hypotensive activity of a plant used in Indian medicine is a sign that, with all its sophisticated means and techniques, the modern pharmaceutical industry still has a lot to learn from the field. Beyond the rupture between "prescientific" and scientific pharmacology, contemporary research is always an extension of traditional, empirical pharmaceutical chemistry, in a history older than ours and common to all the peoples on earth. The pharmacological chemist can certainly pursue the dream of an *a priori* conception of molecules to be synthesized for their pharmaceutical properties, but it is still the case that 60 to 70 percent of medicines today are of natural origin. Here, synthetic chemistry defines itself not only as nature's rival but as an explorer in a labyrinth of tangled natural histories. This kind of work involves the art of intertwining heterogeneous logics and cannot be summarized as optimal calculations of means in service to an end.[5] From this field the chemist takes the active molecules, which he isolates, purifies, copies, and modifies at leisure. But it is also "on the field—on the ailing body"—that medicine designed in a laboratory must operate. Humanity delegates active chemical substances to act not in the aseptic space of a laboratory but in a living labyrinth whose topology varies in time, where partial and circumstantial causalities are so intertwined that they escape any *a priori* intelligibility.

Can we hazard the view here that the identity of chemistry rests on this specific solidarity between the research and the field? This identity, like all the others, is both produced by the present and susceptible to producing a revision of the past, to making interesting what in the past was represented as an obstacle to progress. More precisely, the solidarity emphasized here casts a new light on a longstanding characteristic of chemistry identifiable throughout its history: the ambivalent relationship between the chemist and nature. Master or pupil of nature? Possessor of nature, or possessed by her? The chemist teeters endlessly, without being able to fix upon a single position. Since their distant origins, chemical practices have witnessed both the human power over material processes and the necessity of learning from them, negotiating with them, exploring what they require and imply. Certainly the alchemists tried to achieve impossible dreams of transmutation, perfection, and eternal youth. But the reference to alchemy suggests other resources. In his attempt to accelerate processes that, over eons in the bowels of the earth, brought metals to perfection, the alchemist in his laboratory confronted the problem of time. Later on, chemistry substituted for this finalized but time-oriented notion of nature the image of an industrious if not industrial nature, in which transformations work according to balance sheets, as they do in the laboratory. The great "cycles" of nitrogen, oxygen, and carbon identified the succession of transformations, each one consuming what was produced previously and producing what was consumed afterward, as on an industrial assembly line in perpetual motion, and looping back on itself. But what is a chemical balance if it does not integrate the many time horizons of those different processes that together create global transformation? What are the properties of a substance if one is interested only in deducing them without learning? And how does one learn them if not by painstaking experiments on the circumstances of which they are an integral part or by deciphering the temporal configuration of all the processes at work?

In thus assigning a mixed identity to chemistry—between field and territory, between the empirical and the rational, between a nature full of singularities and the reign of general laws—aren't we supporting a reactionary attitude, ignoring the powers of modern chemistry? Rather than expressing any kind of nostalgia over a long outmoded past, we present this image of chemistry as a sort of regulating idea for its future. One can certainly object that historians should not work in the realm of ideas or of ideal identities, that their role is not to hold out the dazzling utopia of a radiant future. It is not, however, a question here of "predicting" the future but of experimenting on the present with a sensibility sharpened by the analysis of the past. If this sounds too utopian, it is because chemistry's present status as a service science is too often viewed as its fate. Living through the end of illusions long nourished on their aspiration to a territory, chemists sometimes feel they are at an impasse and submit to the current situation instead of living it as a new type of commitment. Is not the role of the historian precisely to resist the "alchemical" identification of the present as the normal and predetermined outcome of a maturation process? To detach the contemporary situation from any image of destiny? To decipher the potentialities written in the present with the help of the long-term perspective, the many possible futures? In sum, emphasizing that history is always open, as we do here, may be the defining feature of historians of scientific disciplines.

GLOSSARY
NOTES
REFERENCES
INDEX

GLOSSARY

Obsolete Terms and Approximate Modern Equivalents

Acid of sea salt (eighteenth century)	Hydrochloric acid
Aqua fortis, or spirit of nitre	Nitric acid
Aqua regia	Mixture of hydrochloric and nitric acid
Cadmia, or calamine	Zinc oxide
Dephlogisticated air	Oxygen
Diana's dove	Pure silver
Feuerluft (eighteenth century)	Oxygen (literally, "fire air")
Fixed air	Carbon dioxide
Inflammable air	Hydrogen
Magnesia alba	Magnesium carbonate
Metallic lime	Metallic oxide
Mirabile salt, or Glauber's salt	Sodium sulfate
Muriatic acid (ca. 1787)	Hydrochloric acid
Oil of vitriol	Concentrated sulfuric acid
Phlogisticated air	Nitrogen
Salt of tartar	Potassium carbonate
Soda	Sodium carbonate
Spirit of salt	Hydrochloric acid
Spirit of vitriol	Sulfuric acid
Verdorbeneluft (eighteenth century)	Nitrogen (literally, "bad air")

NOTES

1. Origins

1. That we find a blacksmith god, Hephaistos, among the great Greek gods is all the more remarkable given that the ancient Greeks held artisanal practices in low esteem. Remember also that copper and tin (which are ingredients of bronze) were the object of an organized system of trade involving the major part of Europe—notably Greece, Transylvania, Spain, England, Denmark—in the first and second centuries, B.C.

2. The description of this procedure, although it does not resort to allegorical language, is impenetrable nevertheless, whether because of a secret method or technical references too different from our own.

3. Identified as Bolos de Mende, who, according to some people lived about 200 years before Jesus Christ, and according to others between the first and second centuries A.D.

4. From hermetically sealed receptacles to secret doctrines and difficult languages, the references to Hermes Trismegistus live on in our language. No doubt the name Hermes Trismegistus derives from *Thoth*, the Egyptian god of writing, whose Greek equivalent was *Hermes*. He illustrates the gnostic suppression of the distinctions between religion and science: he is the original source of secret knowledge on the creation, and acquiring or deciphering this knowledge is seen as a road to salvation.

5. The word *elixir* comes from the Arabic, *al-exir*, as do many other terms— *alchemy, alcohol, alembic, alkali, camphor, talc*. According to Crosland (1978, p. 57), the Arabic words were simply transliterated by translators who were unsure of their exact meaning.

6. See, for example, Halleux (1989), who states that when a medieval alche-

mist attributed to *practica* the function of "certifying," *theoria,* it was meant to confirm that the alchemist-interpreter had understood the author's theory properly. This reference to the author-authority was common in the Middle Ages and not confined to alchemists.

7. The term *bombast* comes from his middle name. *Paracelsus* means "above" or "superior to" Celsus, a famous Roman doctor.

8. According to Paracelsus, during combustion mercury, the active element, escaped, sulfur assured combustibility, and salt was what remained (i.e., the cinders).

9. Christie and Golinski (1982) provide an illustration: an anonymous manuscript from the end of the seventeenth century. In it techniques are understood on the basis of a scheme that points up the difference between separation (solution) and combination (coagulation). The solution, by which a body is resolved into its first principles, can be achieved either by calcination (reduction of the body into lime) or extraction (separation of the body into its fine and coarse particles); and the latter can be done by taking *either* ascending steps, collecting the fine particles (either the dry way, sublimation, or the wet way, which involves the use of solvents), *or* descending steps, collecting the heavy particles (again, either by the hot method or by the cold method). The "exaltation" procedure, so important in alchemical thought because it was an attempt to impart a more intense activity to the body, had been kept by Libavius, but it has disappeared now: it no longer has any meaning in the problematics of separation and combination.

10. It is usually thought that *gas* comes from *geest,* meaning "spirit" in Dutch; but according to other authors, Helmont seemed to be thinking of "chaos" as well, "gas" being neither a substance nor an essence.

11. Pietro Redondi (1987) even suggested making this question the secret motive for the condemnation of Galileo (who, in the *Saggiatore,* had affirmed his atomist convictions).

12. Chemical reversibility is not to be confused with dynamic or thermodynamic reversibility. In this case reversibility does not at all mean that the transformational means are equivalent; it simply refers to the possibility of getting back the product you started with.

2. The Conquest of a Territory

1. Note in a laboratory notebook, February 21, 1773; cited by Grimaux (1888, p. 104).

2. Lavoisier was no exception. Let us mention another example of the "theory of history": in 1912, in order to refute Ernst Mach's philosophy of physics, Max Planck proposed that the "subject of the history" of physics should not be this or that individual physicist but, rather, faith in the unitary intelligibility of the world that prompts "the physicist."

3. For a critique of this idea see Holmes (1989).

4. Note that organic chemistry tests as well as the typical reactions that span a chemical synthesis carry the names of their inventors: Staudinger, Grignard, Diels-Alder, Wurtz-Fittig, etc.

5. This "query," the last and longest in the work, was added with six others to the Latin edition that appeared in 1706: it was then "Query 23." It would become "Query 31" in the English editions of 1717 and 1718 that Newton had augmented with eight new queries.

6. Here is a perverse effect of this fear. Betty Jo Dobbs's book (*The Foundations of Newton's Alchemy or "The Hunting of the Greene Lyon"*), the first detailed study of Newton's practices and ideas on alchemy, was published by the prestigious Cambridge University Press, which is proper for this perfectly serious and erudite book that is not the least bit sensationalistic. For the English, Newtonian alchemy is a respectable field of historical study. It embarrassed the French more and, for their punishment, they will have to buy the translation of Dobbs's book at Guy Tredaniel, Edition de la Maisnie, in the collection "Les Symboles d'Hermes," whose object is to undertake "this study of this great book of the world in which each religion is a page, each myth a sentence and each symbol a word concealing a bit of the primordial Light."

7. Robert Boyle developed the chemical indicators to identify acids and bases. The most common alkaline (basic) salt is tartar salt.

8. See Holmes (1989, pp. 33–59). They are called "middle" salts because the alkaline "salt" is stable and the acid "salt" is volatile.

9. This practice, of which Boyle is seen as the "norm"—namely, in order to belong to the "experimental community," it is necessary to produce and to communicate to colleagues "facts" (news)—began in France without fuss. On Boyle, see S. Shapin, "Pump and Circumstances—Robert Boyle's Literary Technology," *Social Studies of Science*, 14 (1984), pp. 481–520. For the French case, see Christian Licoppe, *Eprouver, rapporter et convaincre: Une etude du compte-rendu experimental a l'epoque moderne*, doctoral thesis, Paris-VII—Denis Diderot, 1994.

10. Homberg also defined the class of "ammoniac salts," which differed from middle salts in that they were the result of the combination of two volatile bodies.

11. The alkali, derived from tartar salt, can also be extracted from plants.

12. As Metzger (1930, p. 196) points out, Boerhaave did not seek to revolutionize chemistry in the name of Newtonian truth; "he cared about truth, not novelty." This search for true knowledge led him to identify points where chemical doctrines had blinded workers, rather than to conceive the plan of Newtonian chemistry.

13. Stahl's major works were not translated into French until much later, at the same time as a great number of Swedish and German books referring to Stahl appeared, especially those dealing with mines and metallurgy. Baron Holbach published the *Traité du soufre* (1767) and the *Traité des sels* (1771), but by then it was too late to "rediscover" Stahl's truth. It was Rouelle's doctrine that represented it.

14. Rouelle, like Juncker, fixed the problem of weight: the lime, which was supposed to be metal, having lost its phlogiston, was heavier than the original metals. He explained this curious fact, like Juncker, by distinguishing between absolute weight (which remained the same, phlogiston being intangible) and specific weight.

15. Published in London in 1727 and translated by Buffon in 1735 under the title of *La Statique des vegetaux et l'analyse de l'air.* The title, like Buffon's introduction, attracted the attention of the French reader to the role of air in Hales's work.

16. Boerhaave also made fire the universal instrument of chemical transformation.

17. Even Lavoisier recognized the power of this discovery, and Kant in the preface to the second edition of the *Critique of Pure Reason* would make the possibility of transforming "metals into calx and calx into metal by adding or taking away something" one of the three cases (with Galileo and Torricelli) from which reason learned that it must force nature to answer its questions.

18. In his article "Chymie" in the *Encyclopédie*, Venel accorded the role of "general and natural" instrument only to fire. As for dissolution in water, he thought it was due to the action of solvents, which had nothing mechanical about it.

19. This book, which would have more than twenty-five editions in various languages, was, after Lémery's *Cours*, the new summation of known facts and processes. It was the result of Boerhaave's teaching, which attracted students from all over Europe.

20. The discriminating reader will notice that lead appears here as a compound.

21. This passage from relative measurement through a set of relationships to the measurement of a quantity would be at the center of the question of equivalents and atoms in the nineteenth century. When Avogadro's number was calculated, atoms would begin to "exist." From the conceptual point of view, this transition is, like the whole of the chemical problematics of the "Newtonian dream," at the heart of Hegel's theory of measurement.

22. Our modern term *base* functions symetrically to the term *acid.* We learn that "acid plus base yields salt." But the name "base" recalls the original asymmetry: alkali gives its "base" to salt.

23. See Holmes (1962) and Kapoor (1965). Note that chemists henceforth recognized compounds that they called "Berthollides," because their proportions were indefinite. An alternative history is imaginable here: what would chemistry have become if "Berthollides" had been dominant in nature?

24. Holmes (1962) and Daumas (1946) refer to Ostwald to pose the problem of the oblivion into which Berthollet's work fell, and emphasize the small number of chemists facing the immense and pressing work of analysis. Ostwald's explanation is remarkable in that it puts the emphasis on the interests of chem-

ists and the return to an older tradition, while more classical historians explained that "chemistry was not yet ripe." It must be said that Ostwald wrote the history of chemistry from the point of view of physical chemistry, which he hoped would attract the interests of chemists (see Chapter 5).

25. Here are the same reactions in modern stoichiometric notation. Using $CaCO_3$ for limestone: $CaCO_3$ yields $CaO + CO_2$; $MgCO_3$ yields $MgO + CO_2$; $Ca(OH)_2 + K_2CO_3$ yields $CaCO_3 + 2KOH$; $MgO + H_2SO_4$ yields $MgSO_4 + H_2O$; $MgSO_4 + K_2CO_3$ yields $MgCO_3 + K_2SO_4$; $CaCO_3 + 2HCl$ yields $CaCl_2 + H_2O + CO_2$; $CaO + 2HCl$ yields $CaCl_2 + H_2O$.

26. Lavoisier, "Mémoire sur la nature du principe qui se combine avec les métaux pendant leur calcination" (Lavoisier, 1864, vol. 2, p. 122).

27. G. L. Buffon, Histoire naturelle, generale et particulière (Paris, 1774), suppl. vol. 1, pp. 1–78.

28. The publication of Lavoisier's complete works was begun in the nineteenth century under the direction of Jean-Baptiste Dumas and then Edouard Grimaux (Oeuvres de Lavoisier, 6 vols., 1862–1879). The publication of the Correspondance is still in progress: four installments have appeared since 1955 (Editions Belin).

29. The polemical use of epistemological declarations, which was also apparent in the controversy between Proust and Berthollet (see above) was pointed out for the first time by Emile Meyerson (1921, vol. 2, pp. 145, 158).

30. Some commentators want to see the hard core of the chemical revolution in this (Berthelot, 1890).

31. This conception is rather close to that offered by J. A. Turgot in the article "Expansibilité" in Diderot's Encyclopédie.

32. A number of chemists objected that perhaps not all acids contained oxygen and that all the gases except oxygen were "a-zooic," hostile to animal life.

33. One could say, using Thomas Kuhn's terminology, that there is an essential tension between divergent and convergent research; see Kuhn (1962, p. 307).

3. A Science of Professors

1. The grateful country expressed its admiration for this great man by making an exception and admitting Madame Berthelot to the Pantheon with him. Until that time the only way for a woman to gain entrance to the Pantheon had been to have a worthy husband at rest there.

2. It became the Annales de chimie et de physique in 1815, and then in 1914 it was divided into the Annales de chimie and the Annales de physique.

3. They were renamed Annalen der Chemie und Pharmacie in 1839, then Justus Liebig's Annalen der Chemie, in 1873.

4. In England the 1815 Apothecaries Act made chemistry obligatory in the medical curriculum. In France chemistry was an essential part of the curriculum

of the Ecole Polytechnique, of the Conservatoire National des Arts et Métiers and then of the Ecole Centrale des Arts et Manufactures.

5. At the University of Giessen, Liebig accepted ten to fifteen students per year around 1830, and about thirty by 1850. In the middle of the nineteenth century the sixteen French faculties of science accepted an average of 111 students per year.

6. The testing center, built in Meudon in June 1793, was given the task of producing hydrogen, balloons, and pilots for the army. The first problem to be solved was to adapt the laboratory procedure developed by Lavoisier and J. B. Meusnier de la Place to large-scale, industrial production. In less than a year, thanks to concerted action by a scientific team consisting of Guyton de Morveau, Berthollet, Fourcroy, Monge, and two officers, all problems were solved, and in June 1794 a balloon observed the battle of Fleurus.

7. The first-year general chemistry course was entrusted to Fourcroy and Vauquelin; a second-year course in animal and vegetable chemistry to Berthollet and Chaptal, and the chemistry of minerals, in the third year, to Guyton (Langins, 1987; Fourcy, 1828).

8. Students who went to study in Paris had a choice between the Polytechnique, the Ecole normale Supérieur, the Ecole des Mines, the Conservatoire National des Arts et Métiers (from 1819 on), the Central School of Arts and Manufacturing (opened in 1829), the Museum of Natural History, the Collège de France, the School of Medicine or the Upper School of Pharmacy, and finally the Faculty of Sciences. In fact, the student would run into the same professors everywhere, for a dozen honored scholars accumulated many chairs and responsibilities. He would nevertheless notice great disparities among establishments, especially between faculties and *grandes écoles*. In the faculties the professors gave more worldly and superficial instruction: brilliant courses decorated with spectacular experiments.

9. In 1838 the School of Mines built a laboratory to instruct students in analytic chemistry; in 1855 the School of Pharmacy developed a practical course in chemistry and toxicology; in the 1860s Edmond Frémy opened a new laboratory at the Museum, which gave free instruction for four years with a tutoring system and about twenty scholarships to aid the students. Finally, in 1868 minister Victor Duruy created the Ecole Pratique des Hautes Etudes to make up for the insufficiencies of university education (Fox and Weisz, 1980).

10. The Ecole Municipale de Physique et de chimie industrielle was inspired by a pioneering model in industrial chemistry, the school in Mulhouse, which the defeat of 1871 had torn away from France.

11. In this section we follow the analysis of Christine Blondel in *Galvani et Volta* Paris: Editions du Seuil, in press).

12. In effect, "oxygenated muriatic acid" (our chlorine) seemed to be obtained from the oxidation of muriatic acid (our HCl), which, like all acids (according to Lavoisier's theory), was itself supposed to contain oxygen.

13. On the other hand, when applied to organic compounds, electrochemical

dualism became unwieldy and cumbersome, and it was attacked because of that in 1836.

14. Berzelius, in his early relationship to atomic theory, illustrates this epistemology (Berzelius, 1819, pp. 18–19).

15. Contrary to widespread opinion, Comte's positivist philosophy was not an obstacle to the acceptance of atomism in France. Auguste Comte declared in his *Cours de philosophie positive* in 1835 that atomism was in harmony with the whole of scientific thought of all types and that it generalized spontaneous and familiar ideas. He thought the transformation of the atomic theory into the theory of equivalents very fortunate and positive, but he hastened to add that it boiled down to "a simple artifice of language, the real thought having remained essentially identical" (Comte, 1830–1842, vol. 1, pp. 611–613).

16. Berzelius sometimes indicated oxygen atoms by two dots above the symbol of the element associated with it. After 1827 he represented two atoms of the same element by a horizontal bar across the symbol of the element. For example, water: 2H + O, was written Ĥ and carbon dioxide Ċ.

17. Using O = 100 for the density of oxygen, Dumas obtained H = 6.24 and N = 88.5. Using the chemical analogy between ammonia (NH_3) and hydrogen phosphate (PH_3), Dumas assumed that the vapor density of phosphorus should be 196, but the experiment gave 392 (Dumas, 1837, pp. 222–227). Such contradictions would be resolved when it was found that all simple bodies do not have the same molecular constitution in the vapor state. While oxygen and nitrogen form diatomic molecules, phosphorus and arsenic form tetratomic molecules, P_4 and As_4. Mercury must be treated as a monatomic molecule.

18. This technique is an example of multidisciplinary scientific achievements. It required Fraunhofer's optics and spectroscope used to test the quality of the glass, Kirchhoff's study of thermal radiation, and the use of the Bunsen burner (introduced in the 1860s).

19. Later, Laurent would edit a *Précis de cristallographie*, which would be read with profit by Louis Pasteur (see below).

20. The distinction between an atom and a molecule was so obvious to him that he simply mentioned it in passing, in a note at the bottom of the page.

21. Tartrate, which forms a deposit in wine barrels, was used to prepare textiles before dyeing. In 1820 Philippe Kestner, an industrial at Mulhouse, noticed a slightly strange tartrate in his barrels. He asked chemists to analyze it. Gay-Lussac in 1826 thought it was a salt formed from a different acid than the tartaric acid. He called it "racemic acid" (from the word for raisin). Berzelius called it "paratartric acid." *Racemic* has become a generic term, characterizing all compounds that are optically inactive mixtures of two optically active isomers.

22. Gerald L. Geison and James A. Secord (1988) point out the difference between Pasteur's presentation of his work in 1860 and the reality of a research enterprise under the direct influence of Laurent in 1848. See also François Dagognet (1967) and Jean Jacques (1992).

23. The chemical family of "aromatic" compounds had been clearly identified by Hofmann in the 1850s.

24. Kekulé's models appeared in memoirs published in the *Bulletin de la Société Chimique*, vol. 3 (1865), pp. 98–110; *Bulletin de l'Académie Royale de Belgique*, vol. 19 (1865), p. 551; *Zeitschrift für Chemie*, vol. 3 (1867), p. 216.

25. On Kolbe's alternative radical theory, his experimental successes in the early 1860s and his nationalism, see the recent biography by Alan J. Rocke (1993).

4. Industrial Expansion

1. The Royal Academy of Sciences' competition set the following task: "Find the simplest and most economical procedure for decomposing sea salt in large quantities, extracting the alkali that is its base in the pure state, separated from any acid or other combination, without having the value of this alkali mineral exceed the price of that imported from the best foreign sources." There were few responses to the competition, whose prize was offered year after year, because candidates who had developed a serious procedure preferred to ask the government for the exclusive privilege of exploiting it (see Smith, 1979, p. 200).

2. In modern notation the two principal reactions at work in the Leblanc procedure can be expressed thus: $2\,NaCl + H_2SO_4 = Na_2SO_4 + 2HCl$ and $Na_2SO_4 + CaCO_3 + 2C = Na_2CO_3 + CaS + 2CO_2$.

3. Nicolas Clément (1799–1841), first assistant to Guyton de Morveau and then professor at the Conservatoire National des Arts et Métiers, operated a factory with his father-in-law, Charles Desormes (1777–1862), who was also a chemist. They got in the habit of signing a single name, Clément Desormes, when they published. They demonstrated by calculation that, contrary to the accepted ideas, the role of saltpeter was not to furnish the oxygen necessary for the combustion of sulfur nor to furnish the heat necessary for the complete combustion producing sulfuric and not sulfurous acid. They suggested that the combustion of the saltpeter produced nitrogen dioxide, which acted as an oxidizing agent on the sulfur dioxide produced by the combustion of sulfur (Smith, 1979, pp. 54–66).

4. Clément and Desormes specified that the nitrogen oxide was not consumed in the reaction but acted as an instrument. In modern notation their interpretation can be written thus: $SO_2 + NO_2 + H_2O = H_2SO_4 + NO$, and $NO + \frac{1}{2}O_2 = NO_2$.

5. Gay-Lussac considered, first of all, that the work was not the only cause of the bad health of the workers—the unhealthy conditions seemed more harmful to him—and, second, that this law was a reflection on the generosity and humanity of entrepreneurs. He claimed that "the factory owner is in the State a true pater familias," to whom honor and protection must be given. After long debate, the law voted in 1841 simply forbad night work for children less than thirteen years old.

6. According to L. F. Haber (1958, p. 55), exports evolved as follows:

Year	Alkali (tons)	Bleaching materials (tons)
1855	53,200	7,500
1867	158,200	15,800
1876	272,800	47,00

7. It is a double decomposition between sea salt and bicarbonate of ammonia; bicarbonate of sodium precipitates out, allowing the reaction to continue, but a chemical equilibrium is established, which is displaced by an excess of salt. The bicarbonate of sodium produced is separated from the "mother-water" by filtration, then calcinated to produce carbonate of sodium. The ammonium chloride is treated with chalk, which absorbs all the chlorine and displaces the ammonia, which can then be recovered:

$$NaCl + CO_3HNH_4 = CO_3HNa + NH_4Cl$$
$$2CO_3HNa = CO_3Na_2 + H_2O + CO_2$$
$$2NH_4Cl + Ca(OH)_2 = CaCl_2 + 2NH_3 + 2H_2O$$

8. England's population rose from 21 million in 1815 to 32 million in 1871.

9. In the latter case, absorption takes place at the roots, thanks to nodes that form by symbiosis with microorganisms called mycorhizomes.

10. For instance, W. Brock wrote in *The Fontana History of Chemistry* (1992, p. 269): "The future of chemistry, as well as industry, after 1865 was, indeed, to lie in structural chemistry at the sign of this hexagon."

11. In the world of products, novelty is always relative, as is shown by the following example. Since 1874 chemists had known of dichlorodiphenyltrichlorethane (DDT), obtained by the German chemist O. Zeidler. Paul Muller, a chemist working for Geigy in Basel, had been looking for a stable and effective insecticide for twenty years when he discovered in 1939 that this substance had interesting insecticidal properties. With a new use, DDT was in fact a new product, and it was immediately inscribed in human history when it arrested an epidemic of typhus that was decimating the American army in Naples in 1943. For this discovery Muller received the Nobel Prize for medicine in 1948.

12. See Morel (1991). The cryolite consists of fluorides of aluminum (AlF_3) and sodium ($3NaF$).

13. The effects of the chemical mobilization were particularly strong in the evolution of the factories of the Société Chimique des Usines du Rhone. Before the war, this was a medium-size enterprise whose principal activity was the production of perfumes for Brazilian carnivals. During the war it converted to the wholesale production of phenol, chlorine, and Ypérite for the French army. From chemistry for fun to chemistry for war, this evolution was matched by an accelerated rate of industrialization in order to achieve mass production (Cayez, 1988, pp. 65–74).

14. Chaim Weizmann's (1874–1952) process, which consisted of fermenting corn with a microorganism, the bacillus *Clostridium acetobutylicum,* to make butanol and acetone, allowed its author to win the sympathies of the British government and to pressure Lord Balfour into signing a declaration on Palestine. So wartime chemistry had another, indirect, impact on world history in the twentieth century.

15. Synthetic gas was made in an experimental factory under a patent on the hydrogenation of finely pulverized charcoal received by Friedrich Bergius, who would receive the Nobel Prize in chemistry with Carl Bosch in 1932.

16. The USSR also joined the race to synthesize rubber, but we have very little information on this subject. In 1932 it synthesized a rubber from a butadiene called "SK", and later one prepared from a chloroprene base, called "Soyprene."

17. A story from the war years says that the word *nylon* stands for "Now you lose, old Nippon."

18. Integrated circuits include aluminum and copper as well as chrome and silicon nitride; for chips, plastics, ceramics, and gold are used.

19. Gallium arsenide not only has a shorter switching time, because of its more mobile electrons, but it has better insulating properties, consumes less energy, and lends itself to the development of optical components and the reception of high frequencies.

5. Dismembering a Territory

1. The term *energetics* is Ostwald's. Duhem did not use it because he rejected all possible substantiation: energy was not, for him, a more real object than atoms. He spoke indiscriminately of the science he was promoting as a generalized "mechanics" or a "thermodynamics."

2. The "Walden inversion" would in fact become, with the work of Emil Fischer, the point of departure for a new development of stereochemistry, integrating not only the spatial characterization of molecular structures but also that of reactive events.

3. Claude Bernard thought so. After Bernard's death in 1878, Marcellin Berthelot would publish a text to this effect, which would really annoy Pasteur. The alcoholic "ferment," which Bernard supposed could act independently of the vital activity of microorganisms, was only a hypothesis, wrote Pasteur, and "hypotheses like that, pardon the vulgarity of the expression, we brew by the bucket full in the laboratories . . . Between Mr. Berthelot and me there is this difference: I never let this kind of hypothesis see the light of day until I am sure it is true and it allows us to go forward. Mr. Berthelot publishes them." To argue against the existence of a soluble alcoholic ferment (i.e., one separable from microorganisms), Pasteur used time as his ally: "How could M. Berthelot not feel that time is the only judge in this matter, and the sovereign judge? How could he not recognize that I don't have to complain about the verdict of time?

Doesn't he see the fruitfulness of the conclusions of my previous studies growing every day?" (cited in Jacques, 1987, pp. 157–158).

4. Or to "colloids," according to the expression introduced by Thomas Graham in 1861 to distinguish them from "crystalloid" substances susceptible only to states of static equilibrium, like crystals.

5. Engels (1960, pp. 195–196; italicized in the original). Unlike Liebig, Engels supported the distant possibility that these albuminoid bodies could be prepared artificially and would then exhibit vital phenomena unmistakably, no matter how weak and ephemeral they might be, since the organisms that we know are the product of thousands of years of evolution.

6. Thus, remarked Ostwald, "the activity of aluminum chloride in Friedel's and Crafts's reaction could be compared to *Tischlein-deck-dich* [Table, set yourself] in the old German story, because it facilitated so greatly the formation of bodies that would be extremely difficult to obtain without it" (Ostwald, [1906] 1909, p. 286).

7. The physics of states far from equilibrium, as we shall see, retrieved the notion of permanent exchange as a condition of existence, but it was not a question of the existence of substances, properly speaking, but of global regimes of activity involving all the physical processes and chemical reactions of the system.

8. The priority of Thomsen's work, undertaken between 1850 and 1860, over that of Berthelot, who became interested in thermochemistry only in 1864, was one of Pierre Duhem's warhorses in his long struggle against his persecutor. In the eyes of Duhem, Thomsen also had the advantage of having made known the work of Guldberg and Waage (see below) and of having abandoned as early as 1869 the principle on which the priority dispute turned.

9. That is to say, there is no "principle of inertia" in chemistry, and forces are defined as they were in mechanics before Galileo. To this difference between definitions of the force corresponds a phenomenological difference: whereas a mechanical system oscillates around a state of equilibrium (damped oscillations in case of friction), a chemical system generally goes to equilibrium in a monotonous way. When the forces cancel each other out, all chemical transformation stops.

10. This science did develop, in the twentieth century. It is based on the notion of the "activated complex," a transitory situation in which molecules in contact form an unstable being, which could give back either the original molecules or new ones. It was a hybrid theory associating theoretical elements (the stability of structures calculated from quantum chemistry), empirical elements (thermodynamic quantities, such as heat given off or absorbed), and experimental data (by this time new experimental techniques could identify directly many activated complexes in spite of their short life).

11. That explains some mechanisms of catalysis: when the same molecule constitutes a reagent for an intermediary step and a reaction product for another, it disappears from the balance sheet and seems not to have participated in the reaction except as a catalyst.

12. The system dilates as it is heated at constant temperature; it contracts as it is cooled at constant temperature. Its temperature increases as it is compressed without exchange of heat; its temperature decreases as it is expanded without exchange of heat. There are, in this theory, two "isotherms" and two "adiabatics."

13. The measurement of equilibrium displacement was given in terms of the work required to "displace" the chemical composition in a reversible way, by adding or subtracting heat at constant pressure, or else (when there was a gaseous phase) by compression or dilation at constant temperature. The higher the temperature, the more the equilibrium favored bodies whose combination was endothermic (heat-absorbent). Van't Hoff showed that his law coincided in one case with Berthelot's principle of maximum work: it was the case in which reactions took place at absolute zero. This allowed French chemists to continue to teach Berthelot's principle for quite a while, by discretely introducing this "slight" restriction.

14. The chemical potential proposed by Duhem systematized different definitions of the thermodynamic functions in chemistry elaborated by F. J. D. Massieu (in 1869), Max Planck (1869), Josiah Willard Gibbs (1876), and Hermann Helmholtz (1882).

15. On this subject, see a defense of mechanism against mathematical abstraction by Abel Rey (1923), whom Lenin called "confusionist Rey" in *Materialism and Empiricism.* The original text was written in 1905, and the 1923 edition announced, on the basis of relativity and quanta, the triumph of mechanism, which had been in danger twenty years earlier.

16. In this description we have neglected the γ rays of light. Each product named obviously has an atomic weight equal to the preceding one if its production has been accompanied by a β emission, and a weight lower by two units if it was accompanied by an α emission.

17. As a consequence, starting in the 1910s Marie Curie's lab had a real "capital"—an enormous quantity of this precious, rare, and costly radium, which other laboratories desperately lacked. Frederic Soddy (1926, pp. 29–31) complained of this fact in veiled terms.

18. In 1934, when artificial radioactivity was discovered, Irène Curie, the daughter of Pierre and Marie, used the art of the radiochemist to succeed in identifying in three minutes, before its disappearance, isotope 30 of the phosphorus produced by the irradiation of aluminum.

19. According to the correspondence principle, the quantum description should rejoin the laws of classical physics in the limit of high quantum numbers—i.e., when the quantum of energy becomes small compared with the energy of the orbital states; these states can then be considered to form a continuum (see Max Jammer, 1966).

20. This situation, which has been stable for more than sixty years, is not necessarily definitive. Ilya Prigogine thus pleaded for a generalization of quantum mechanics to systems of the "chaotic" type, such that Schrödinger's equa-

tion would become no more than a singular limiting case, while the general mathematical structure would be that of kinetic equations: probabilistic and realistic.

21. Cited in Palmer (1965, p. 160). The homopolar bond is the covalent bond that unites two identical atoms (same poles) to form a molecule. While the existence of simple, diatomic bodies seemed an *ad hoc* hypothesis to Avogadro's critics, the bond that unites the two atoms of the hydrogen molecule has since become the prototype of the covalent bond.

22. The crusade for a materialist chemistry was accompanied by a rewriting of the history of structural chemistry, in which the Russian chemist Alexander Butlerov, until then (unjustly) ignored by historians of chemistry, suddenly had a glory equal to that of Kekulé attributed to him. For the Soviets, "Machian" idealism (derived from Ernst Mach's pragmatic philosophy) could be traced back to Kekulé, and the true materialist tradition to Butlerov. In fact, Butlerov, like Kekulé, wanted the chemical formula to represent the facts in a convenient way (Graham, 1974, pp. 304–305).

23. This atypical situation, like the possibility of his own work, was often linked by Prigogine to the relatively eccentric character of Brussels, compared to the great priorities organizing the international scientific competition.

24. Certain reactions can correspond to a "decrease" of entropy; it is enough that the entropic balance of the coupled reactions as a whole be positive.

25. The notion of "negative entropy" of a system "consuming" the entropy furnished by the environment, is a purely formal solution to the problem. The thermodynamic question is not only to escape the interdict of the second principle but to use it as a principle of evolution and to find out what types of behavior a system will actually adopt far from equilibrium. Today the same kind of difference separates Henri Atlan's "order by noise" and Ilya Prigogine's "order by fluctuations."

26. The German chemist Manfred Eigen and his collaborators study models of "chemical histories" in order to understand the production of some molecular structures in preference to others. The starting point is the fact that certain polymers, by their own structure, are capable of catalyzing their own copy. If parameters characterizing the rate and exactitude of the copy are attributed to each member of such a polymer, one can follow a "selective competition" between polymers. A more complicated model, which crosses two types of polymers with distinct roles, ends with the exclusive domination by a kind of proto-genetic code: a specific association (hypercycle) of polymers, which is stable in relationship to "mutants" that continue to produce unfaithful copies. See Eigen and Schuster (1979).

Epilogue

1. The *New Scientist* of February 22, 1992, announced on p. 19 the first "computer-invented synthesis" by a computer that churned out seventy-two

reactions for synthesizing butadiene, two of which chemists had never thought of (one of them produced a return of 95 percent).

2. See on this subject Jacques Monod's hymn to "molecular cybernetics" in *Le Hasard et la Nécessité*.

3. Let us remark that at the "fundamental" level of quantum representation of the atom, this could also prove unstable as a result of the new experimental devices allowing us to "visualize" atoms or events. See Von Baeyer (1993).

4. It is because of traces of rare-earth oxides in tobacco that a piece of sugar wrapped in cigarette ashes burns.

5. See Ourisson (1991). Whereas the biologist and the pharmacologist decipher in the secondary metabolites synthesized by plants a multitude of functions, poisons, bactericides, odors, dyes, etc., the synthetic chemist reads the relative unity of the great biosynthetic pathways: some common precursors and the divergence of synthetic paths. Not only does this "universal" biosynthetic organization introduce a "chemotaxonomic" order (which duplicates the biological classification of species), it also allows us to stage the chemical landscape of opportunities and constraints as the playground for Darwinian selection.

REFERENCES

Aftalion, F. 1987. *Histoire de la chimie.* Paris: Masson.

Anastasi, A. 1884. *Nicolas Leblanc, sa vie, ses travaux, et l'histoire de la soude artificielle.* Paris: Hachette.

Anderson, W. C. 1984. *Between the Library and the Laboratory: The Language of Chemistry in Eighteenth-Century France.* Baltimore: Johns Hopkins University Press.

Avogadro, A. [1811] 1991. "Essai d'une manière de déterminer les masses relatives des molécules élémentaires des corps et les proportions selon lesquelles elles entrent dans ces combinaisons." *Journal de physique, de chimie et d'histoire naturelle,* 73, pp. 58–79. Reprinted in *Les Atomes, Une anthologie historique.* Paris: Presses Pocket, 1991.

Bachelard, G. [1930] 1973. *Le Pluralisme cohérent de la chimie moderne.* Paris: Vrin.

Balard, A. J. 1862. *Rapport sur les industries chimiques à l'Exposition universelle de 1862.* Paris: Hachette.

Baranger, P. 1956. "L'âge chimique." In *Les Grandes Découvertes du XXe siècle,* ed. Le Prince-Ringuet, pp. 173–204. Paris: Larousse.

Baudrimont, A. E. 1833. *Introduction à l'étude de la chimie par la théorie atomique.* Paris.

Beer, J. J. 1959. *The Emergence of the German Dye Industry.* Urbana: University of Illinois Press.

Ben-David, J. 1984. *The Scientist's Role in Society: A Comparative Study.* Chicago: University of Chicago Press.

Bensaude-Vincent, B. 1982. "L'éther, élément chimique: un essai malheureux de Mendeleev?" *British Journal for the History of Science,* 15, pp. 183–188.

———— 1989. "Lavoisier: une révolution scientifique." In *Éléments d'histoire des sciences,* ed. M. Serres, pp. 363–386. Paris: Bordas.

———— 1989. "Mendeleev." In *Éléments d'histoire des sciences,* ed. M. Serres, pp. 447–468. Paris: Bordas.

———— 1990. "Karlsruhe, septembre 1860: l'atome en congrès." *Relations internationales,* 62, pp. 149–169.

Beretta, M. 1988. "T. O. Bergman and the Definition of Chemistry." *Lychnos,* pp. 37–67.

Bernard, C. 1865. *Introduction à la médecine expérimentale.* Paris: Baillère.

Berthelot, M. 1860. *Chimie organique fondée sur la synthèse,* 2 vols. Paris: Mallet Bachelier.

———— 1876. *La Synthèse chimique.* Paris: Félix Alcan.

———— 1890. *La Révolution chimique: Lavoisier.* Paris: Félix Alcan.

Berzelius, J. J. 1819. *Essai sur la théorie des proportions chimiques et sur l'influence chimique de l'électricité.* Paris.

Beyewets, A. S. P. 1986. *La Chimie des matières colorantes artificielles.* Paris: Masson.

Blackley, D. C. 1983. *Synthetic Rubbers: Their Chemistry and Technology.* London and New York: Applied Science Publishers.

Boas, M. 1958. *Robert Boyle and Seventeenth-Century Chemistry.* Cambridge: Cambridge University Press.

Bockris, J. 1991. "Cold Fusion II: the Story Continues." *New Scientist,* January 19, pp. 50–53.

Boussingault, J.-B., and J.-B. Dumas. [1842] 1972. *Essai de statique chimique des êtres organisés.* Brussels.

Boyle, R. 1992. *Sceptical Chymist.* Kila, MT: Kessinger Publishing Co.

Braudel, F. 1977. *The Mediterranean and the Mediterranean World in the Age of Philip II,* 2 vols., 2nd ed., trans. S. Reynolds. New York: HarperCollins.

Brock, W. B. 1985. *From Protyle to Proton: William Prout and the Nature of Matter.* Bristol and Boston: Adam Hilger Ltd.

Brooke, J. H. 1968. "Wöhler's Urea and the Vital Force—A Verdict from the Chemists." *Ambix,* 15, pp. 84–114.

———— 1971. "Organic Chemistry and the Unification of Chemistry: A Reappraisal." *British Journal for the History of Science,* 5, pp. 369–392.

———— 1981. "Avogadro's Hypothesis and Its Fate: A Case Study in the Failure of Case-Studies." *History of Science,* 19, pp. 235–273.

Brouzeng, P. 1987. *Duhem: Science et providence.* Paris: Belin.

Bud, R. F. 1992. *The Uses of Life: A History of Biotechnology.* Cambridge: Cambridge University Press.

Bud, R. F., and G. K. Roberts. 1984. *Science versus Practice: Chemistry in Victorian Britain.* Manchester: Manchester University Press.

Butterfield, H. 1965. *The Origins of Modern Science, 1300–1800,* rev. ed. New York: Free Press.

Bynum, W. F., E. J. Browne, and R. Porter. 1981. *Dictionary of the History of Science.* Princeton, N.J.: Princeton University Press.

Cardwell, D. S. L. 1968. *John Dalton and the Progress of Science*. Manchester: Manchester University Press.

———— 1971. *From Watt to Clausius: The Rise of Thermodynamics in the Early Industrial Age*. London: Heinemann.

Caro, P. 1991. "La recherche en chimie: évolution et perspectives." *La Chimie, ses industries et ses hommes. Culture technique*, no. 23, CRCT, pp. 209–217.

Cartwright, N. 1983. *How the Laws of Physics Lie*. Oxford: Clarendon Press.

Carusi, P. 1990. "Alchemy." In *A History of Chemistry*, ed. A. Di Meo, pp. 33–71. Venice: Marsilio Editori.

Cayez, P. 1988. *Rhône-Poulenc 1895–1975*. Paris: Armand Colin/Masson.

Chargaff, E. 1980. *Heraclitean Fire*. New York: Warner Books.

Chevreul, M. E. 1823. *Recherches chimiques sur les corps gras d'origine animale*. Paris: Lavrault.

———— 1824. *Considérations générales sur l'analyse organique et ses applications*. Paris: Lavrault.

Christie, J. R., and J. V. Golinski. 1982. "The Spreading of the Word: New Directions in the Historiography of Chemistry, 1600–1800." *History of Science*, 20, pp. 235–266.

Christophe, R. 1989. *Guide des sources concernant la formation des ouvriers des métiers et industries chimiques 1750–1870*. Paris: Cité des sciences et de l'industrie.

Cohendet, P., J. M. Ledoux, and E. Zuscovitch. 1987. *Les Matériaux nouveaux. Dynamique économique et stratégie européenne*. Paris: Economica.

Cole, W. A. 1988. *Chemical Literature 1700–1860*. London.

Collectif. 1982. *Les Années plastiques*. Paris: Alternatives/Cité des sciences et de l'industrie.

Colnort-Bodet, S. 1986. *Du pneuma aux grades et à l'universel ou la maturation de la notion de quantité chez les thérapeutes et les techniciens*. Doctoral thesis, Paris-IV.

Compain, J.-C. 1992. "Les travaux de Le Bel et Van't Hoff de 1874 et notre enseignement." *Bulletin de l'Union des physiciens*, 86, pp. 285–310.

Comte, A. –1975. *Cours de philosophie positive*, 2 vols. Paris: Edition Hermann.

Crosland, M. P. [1962] 1978. *Historical Studies in the Language of Chemistry*. New York: Dover.

———— 1971. *The Society of Arcueil: A View of the French Science at the Time of Napoleon I*. London: Heinemann.

———— 1975. "The Development of a Professional Career in Science in France." *Minerva*, 13, pp. 38–57.

———— ed. 1975. *The Emergence of Science in Western Europe*. London: Macmillan Press.

———— 1978. *Gay-Lussac, Scientist and Bourgeois*. Cambridge: Cambridge University Press.

Culture technique. 1984. "Les Ingénieurs." In *Culture technique*, no. 12. Paris: CRCT.

Culture technique. 1991. "La Chimie, ses industries et ses hommes." In *Culture technique,* no. 23. Paris: CRCT.

Dalton, J. [1808] 1964. *A New System of Chemical Philosophy.* New York: Philosophical Library.

Dagognet, F. 1967. *Méthodes et doctrines dans l'oeuvre de Pasteur.* Paris: Presses universitaires de France.

———— 1969. *Tableaux et langages de la chimie.* Paris: Vrin.

Daumas, M. 1946. *L'Acte chimique. Essai sur l'histoire de la philosophie chimique.* Brussels: Editions du Sablon.

Debus, A. G. 1992. *The French Paracelsians: The Chemical Challenge to Medical and Scientific Tradition in Early Modern France.* Cambridge: Cambridge University Press.

De Jouy, E. 1822. *L'Ermite en province ou Observations sur les mœurs et les usages français au commencement du XIXe siècle,* 3d ed. Paris.

Dijksterhuis, E. J. [1950] 1986. *The Mechanization of the World Picture.* Princeton: Princeton University Press.

Di Meo, A., ed. 1990. *Storia della chimica: dalla ceramica del neolitico all'eta della plastica,* 2d ed. Venice: Marsilio Editori.

Dobbs, B. J. T. 1975. *The Foundations of Newton's Alchemy, or "The Hunting of the Greene Lyon."* Cambridge: Cambridge University Press.

———— 1982. "Newton's Alchemy and His Theory of Matter." *Isis,* 73, pp. 511–528.

Donnelly, J. 1991. "Industrial Recruitment of Chemistry Students from English Universities: A Reevaluation of Its Early Importance." *British Journal for the History of Science,* 24, pp. 3–20.

Donovan, A. 1975. *Philosophical Chemistry in the Scottish Enlightenment.* Edinburgh: Edinburgh University Press.

———— ed. 1988. "The Chemical Revolution: Essays in Reinterpretation." *Osiris,* 2nd ser., vol. 4.

D'Or, L. [No date.] "Notice biographique sur Ernest Solvay." In *Florilège des sciences en Belgique,* pp. 385–406. Brussels: Académie royale de Belgique.

Drouot, M., A. Rohmer, and N. Stoskopf. 1991. *La Fabrique des produits chimiques, Thann et Mulhouse, histoire d'une entreprise de 1808 à nos jours.* Strasbourg.

Duhem, P. [1902] 1985. *Le Mixte et la combinaison chimique.* Corpus des oeuvres de philosophie en langue française. Paris: Fayard.

Dumas, J.-B. [1837] 1972. *Leçons sur la philosophie chimique.* Brussels: Editions Culture et Civilisation.

Eigen, M., and P. Schuster. 1979. *The Hypercycle.* Berlin: Springer-Verlag.

Eisenstein, E. 1983. *The Printing Revolution in Early Modern Europe.* Cambridge: Cambridge University Press.

Eklund, J. 1975. *The Incompleat Chymist, Being an Essay on the Eighteenth-Century Chemist in his Laboratory, with a Dictionary of Obsolete Chemical Terms of the Period.* Washington, D.C.: Smithsonian Institution Press.

Elkana, Y. 1974. *The Discovery of the Conservation of Energy*. London: Hutchinson.

Emptoz, G. 1991. "Des produits chimiques très recherchés: les acides gras pour la fabrication des bougies stéariques." *Culture technique*, 23, pp. 32–45.

Engels, F. 1960. *Dialectics of Nature*, trans. and ed. C. Dutt. New York: International Publishers.

Fabre, J.-H. 1925. *Souvenirs entomologiques*, vol. 10. Paris: Delagrave.

Finlay, M. R. 1991. "The Rehabilitation of an Agricultural Chemist: Justus von Liebig and the Seventh Edition." *Ambix*, 38, pp. 155–167.

Falconer, I. 1987. "Corpuscules, Electrons and Cathode Rays." *British Journal for the History of Science*, 20, pp. 241–276.

Fischer, N. 1982. "Avogadro, the Chemists and Historians of Chemistry." *History of Science*, 20, pp. 77–102, 212–231.

Fischer, W. 1978. "The Role of Science and Technology in the Economic Development of Germany." In *Science, Technology and Economic Development*, ed. W. Beranek and G. Ranis, pp. 73–113. New York: Praeger.

Fourcy, A. [1828] 1987. *Histoire de l'École polytechnique*. Paris: Belin.

Fox, R. 1973. "Scientific Enterprise and the Patronage of Research in France, 1800–1870." *Minerva*, 11, pp. 442–473.

———— 1984. "Presidential Address: Science, Industry, and the Social Order in Mulhouse, 1798–1871." *British Journal for the History of Science*, 17, pp. 127–165.

Fox, R., and G. Weisz. 1980. *The Organization of Science and Technology in France, 1808–1914*. Cambridge: Cambridge University Press.

French, S. J. 1941. *Torch and Crucible: The Life and Death of Antoine Lavoisier*. Princeton: Princeton University Press.

Freund, I. [1904] 1968. *The Study of Chemical Composition: An Account of Its Method and Historical Development*, 2nd ed. New York: Dover Publications.

Fricke, M. 1976. "The Rejection of Avogadro's Hypothesis." In *Method and Appraisal in the Physical Sciences*, ed. C. Howson. Cambridge: Cambridge University Press.

Fruton, J. S. 1990. *Contrasts in Scientific Style: Research Groups in the Chemical and Biochemical Sciences*. Philadelphia: American Philosophical Society.

Gaudin, M. A. 1833. *L'Architecture du monde des atomes*. Paris.

Geison, G. L., and J. Secord. 1988. "Pasteur and the Process of Discovery: The Case of Optical Isomerism." *Isis*, 79, pp. 6–36.

Gerhardt, C. 1853–1856. *Traité de chimie organique*, 4 vols. Paris.

Gillispie, C. C. 1957. "The Discovery of the Leblanc Process." *Isis*, 48, pp. 152–170.

———— 1970–1980. *Dictionary of Scientific Biography*. New York: Scribners and Sons.

Gittings, F. 1981. *Dictionary of Occult, Hermetic and Alchemical Sigils*. Boston: Routledge and Kegan Paul.

Glansdorff, P., and I. Prigogine. 1971. *Structure, stabilité et fluctuations*. Paris: Masson.

Goldbeter, A. 1990. *Rythmes et chaos dans les systèmes biochimiques et cellulaires.* Paris: Masson.

Golinski, J. W. 1992. *Science as Public Culture: Chemistry and the Enlightenment in Britain, 1760–1820.* Cambridge: Cambridge University Press.

Graham, L. R. 1974. *Science and Philosophy in the Soviet Union.* New York: Vintage Books.

Grandmougin, R. 1917. *L'Enseignement de la chimie industrielle en France.* Paris: Dunod.

Grimaux, E. [1888] 1992. *Lavoisier d'après sa correspondance, ses manuscrits.* Paris: Jean Gabay.

Guédon, J.-C. 1971. "Le lieu de la chimie dans l'*Encyclopédie*." *XIIIe Congrès d'histoire des sciences,* 7, pp. 80–86.

———— 1983. "From Unit Operations to Unit Processes." In *Chemistry and Modern Society: Historical Essays in Honour of Aaron J. Idhe,* ed. J. Parascandola and J. W. Whorton. Washington, DC: American Chemical Society.

Guerlac, H. 1961. *Lavoisier, The Crucial Year: The Background and Origin of His First Experiments on Combustion in 1772.* Ithaca, NY: Cornell University Press.

———— 1975. *Antoine-Laurent Lavoisier: Chemist and Revolutionary.* New York: Scribners.

Guyton de Morveau, L. B. 1786. *Dictionnaire de chimie de l'encyclopédie méthodique,* vol. 1. Paris.

Haber, L. F. 1958. *The Chemical Industry during the Nineteenth Century.* Oxford: Clarendon Press.

———— 1968. *The Poisonous Cloud: Chemical Warfare in the First World War.* Oxford: Clarendon.

Hafner, K. 1979. "August Kekulé, the Architect of Chemistry: Commemorating the 150th Anniversary of His Birth." *Angewandte Chemie,* 18, p. 641–706.

Halleux, R. 1989. "Recettes d'artisan, recettes d'alchimiste." *Artes Mechanicae en Europe médiévale,* ed. R. Jansen-Siebened. Archives et bibliothèques de Belgique, special no. 34. Brussels.

Hannaway, O. 1975. *The Chemists and the Word: The Didactic Origins of Chemistry.* Baltimore and London: Johns Hopkins University Press.

Hoefer, F. 1840–1842. *Histoire de la chimie.* Paris: Firmin Didot.

Hoffmann, R. 1991. "Apologie de la synthèse." *Alliage,* no. 9, pp. 65–76.

Holmes, F. L. 1962. "From Elective Affinities to Chemical Equilibria: Berthollet's Law of Mass Action." *Chymia,* 8, pp. 105–145.

Holmes, L. F. 1985. *Lavoisier and the Chemistry of Life: An Exploration of Scientific Creativity.* Madison: University of Wisconsin Press.

———— 1989. *Eighteenth-Century Chemistry as an Investigative Enterprise.* Berkeley: University of California Press.

Homburg, E. 1983. "The Influence of Demand on the Emergence of the Dye Industry: The Roles of Chemists and Colourists." *Journal of the Society of Dyers and Colourists,* 99, pp. 325–333.

Hornix, W. J. 1992. "From Process to Plant: Innovation in the Early Artificial Dye Industry." *British Journal for the History of Science*, 25, pp. 65–90.

Hufbauer, K. 1982. *The Formation of the German Chemical Community (1720–1795)*. Berkeley: University of California Press.

Jacques, J. 1954. "La thèse de doctorat d'Auguste Laurent et la théorie des combinaisons organiques (1836)." *Bulletin de la Société chimique de France*, pp. 31–39.

———— 1987. *Marcellin Berthelot. Autopsie d'un mythe*. Paris: Belin.

———— 1992. *La Molécule et son double*. Paris: Hachette/CSI.

Jammer, M. 1966. *The Conceptual Development of Quantum Mechanics*. New York: McGraw-Hill.

Johnson, L. 1991. "The First Enzyme Picture Show." *New Scientist*, December 14, pp. 30–33.

Judson, H. F. 1979. *The Eighth Day of Creation: The Makers of the Revolution in Biology*. New York: Simon and Schuster.

Kapoor, S. J. 1965. "Berthollet, Proust and Proportions." *Chymia*, 10, pp. 53–110.

Kekulé, A. [1861] 1866. *Lechbuch der Organischen Chemie*, vol. 1. Erlangen.

———— [1861] 1882. *Lechbuch der Organischen Chemie*, vol. 2. Stuttgart.

———— 1864. "Sur l'atomicité des éléments." *Comptes rendus de l'Académie des sciences*, 58, pp. 510–514.

Kirwann, R. 1780. *Notes sur C. W. Scheele: Chemical Observations and Experiments on Air and Fire*. London.

Klosterman, L. J. 1985. "A Research School of Chemistry in the Nineteenth Century: J. B. Dumas and His Sudents." *Annals of Science*, 42, pp. 1–40.

Knight, D. M. 1970. *Atoms and Elements*, 2nd ed. London: Hutchinson.

———— 1992. *Ideas in Chemistry: A History of the Science*. London: Athlone Press Ltd.

Kohler, R. E. 1975. "G. N. Lewis's Views on the Bond Theory." *British Journal for the History of Science*, 8, pp. 233–239.

Kopp, H. 1843–1847. *Geschichte der Chemie*. Braunschveig.

Kuhn, T. S. 1952. "Robert Boyle and Structural Chemistry in the Seventeenth Century." *Isis*, 43, pp. 12–16.

———— 1962. *The Structure of Scientific Revolutions*. Chicago: University of Chicago Press.

Ladenburg, A. [1879] 1911. *Histoire du développement de la chimie depuis Lavoisier jusqu'à nos jours*, 2nd ed. Paris: Hermann.

Langins, J. 1987. *La République avait besoin de savants*. Paris: Belin.

Latour, B. 1989. *Science in Action*. Cambridge, MA: Harvard University Press.

Laurent, A. 1836. "Théorie des combinaisons organiques." *Annales de chimie*, 61, pp. 125–146.

Lavoisier, A. L. 1862–1864. *Oeuvres*, 2 vols. I and II. Paris: Imprimerie impériale.

Le Bel, A. 1874. "Sur les relations qui existent entre les formules atomiques des corps organiques et le pouvoir rotatoire de leurs dissolutions." *Bulletin de la Société chimique de Paris*, November. [Reprinted by J. H. van't Hoff, 1874.]

Le Bras, J. 1969. *Le Caoutchouc.* Paris: Presses universitaires de France.

Le Chatelier, H. 1925. *Science et Industrie.* Paris: Flammarion.

Leprieur, F. 1977. *Les Conditions de la constitution d'une discipline scientifique: la chimie organique en France (1830–1880).* Thesis, Paris-I.

Leprieur, F., and P. Papon. 1979. "Synthetic Dyestuff: The Relations between Academic Chemistry and the Chemical Industry in Nineteenth-Century France." *Minerva,* 17.

Leroux-Calas, M. 1991. "La recherche au service de la production d'aluminium." In *Histoire technique de la production de l'aluminium,* ed. P. Morel, pp. 87–107. Grenoble: Presses universitaires de Grenoble.

Levere, T. H. 1971. *Affinity and Matter: Elements of Chemical Philosophy, 1788–1865.* Oxford: Clarendon Press.

Levi, P. 1984. *The Periodic Table,* trans. R. Rosenthal. New York: Schocken.

Liebig, J. [1841] 1842. *Chmistry in Its Application to Agriculture and Physiology,* 3d American ed., ed. L. Playfair. Cambridge, MA: John Owen.

——— 1845. *Lettres sur la chimie.* Paris: Librairie Baillère.

——— [1863] 1894. *Lord Bacon.* Paris: Librairie Baillière.

Lindsay, J. 1970. *Origins of Alchemy in Graeco-Roman Egypt.* London: Muller.

Macleod, R. 1993. "The Chemists Go to War: The Mobilization of Civilian Chemists and the British War Effort, 1916–1918." *Annals of Science,* 50, pp. 445–481.

Massain, R. 1952. *Chimie et chimistes.* Paris: Editions Magnard.

Mauskopf, S. M. 1976. "Crystals and Compounds." *Transactions of the American Philosophical Society,* 66, part 3, pp. 5–82.

Mehra, J. 1975. *The Solvay Conferences on Physics.* Boston: Reidel.

Meinel, C. 1983. "Theory or Practice? The Eighteenth-Century Debate on the Scientific Status of Chemistry." *Ambix,* 30, pp. 121–132.

——— 1988. "Early Seventeenth-Century Atomism: Theory, Epistemology, and the Insufficiency of Experiment." *Isis,* 79, pp. 68–103.

Mendeleev, D. [1869–1871] 1905. *Principles of Chemistry,* 2nd English ed., trans. from the Russian (6th ed.) by G. Kamensky. New York: P. F. Collier.

——— [1871] 1879. "La loi périodique des éléments chimiques." *Le Moniteur scientifique,* 21, pp. 691–737.

——— 1904. *An Attempt Towards a Chemical Conception of Ether.* London, New York, Bombay.

Metzger, H. [1918] 1969. *La Genèse de la science des cristaux.* Paris: Blanchard.

——— [1923] 1969. *Les Doctrines chimiques en France du début du XVIIe à la fin du XVIIIe siècle.* Paris: Blanchard.

——— [1930] 1974. *Newton, Stahl, Boerhave et la doctrine chimique.* Paris: Blanchard.

Meyerson, E. 1921. *De l'explication dans les sciences,* 2 vols. Paris: Félix Alcan.

Meyer-Thurow, G. 1982. "The Industrialization of Invention: A Case Study in Chemical Industry." *Isis,* 73, pp. 363–381.

Moore, J. F. 1939. *A History of Chemistry.* New York: McGraw-Hill.

Morazé, C. 1986. *Les Origines sacrées des sciences modernes*. Paris: Fayard.

Morel, P., ed. 1991. *Histoire technique de la production d'aluminium*. Grenoble: Presses universitaires de Grenoble.

Morrell, J. B. 1972. "The Chemist Breeders: The Research Schools of Justus Liebig and Thomas Thomson." *Ambix*, 19, pp. 1–43.

Morrell, J. B., and A. Thackray. 1981. *Gentlemen of Science*. Oxford: Clarendon Press.

Morsel, H. 1991. "Introduction." In *Histoire technique de la production d'aluminium*, ed. P. Morel, pp. 1–20. Grenoble: Presses universitaires de Grenoble.

Munday, P. 1991. "Liebig's Metamorphosis: From Organic Chemistry to the Chemistry of Agriculture." *Ambix*, 38, pp. 133–154.

Needham, J. 1970. *Clerks and Craftsmen in China and the West*. Cambridge: Cambridge University Press.

Newman, W. R. 1994. *Gehennical Fire: The Lives of George Starkey, an American Alchemist in the Scientific Revolution*. Cambridge, MA: Harvard University Press.

Newton, I. [1717] 1952. *Opticks: or A Treatise of the Reflections, Refractions, Inflections & Colours of Light*. New York: Dover.

Nye, M.-J. 1972. *Molecular Reality: A Perspective of the Scientific World of Jean Perrin*. London: MacDonald.

———— 1976. "The Nineteenth-Century Atomic Debates and the Dilemma of an Indifferent Hypothesis." *Studies in History and Philosophy of Science*, 7, pp. 245–268.

———— 1983. *The Question of the Atom: From the Karlsruhe Congress to the Solvay Conference, 1860–1911*. Los Angeles: Tomash.

———— 1986. *Science in the Provinces: Scientific Communities and Provincial Leadership in France, 1860–1930*. Berkeley: University of California Press.

Ostwald, W. [1906] 1909. *L'Évolution d'une science. La chimie*. Paris: Flammarion.

Ourisson, G. 1991. "Ordre ou désordre?" *La Chimie, ses industries et ses hommes. Culture technique*, no. 23, CRCT, pp. 141–149.

Palmer, W. G. 1965. *A History of the Concept of Valence to 1930*. Cambridge: Cambridge University Press.

Partington, J. R. 1961–1964. *A History of Chemistry*, 4 vols. London: Macmillan.

Pasteur, L. [1860] 1986. "Recherches sur la dissymétrie moléculaire des produits organiques naturels." *Leçons de chimie professées en 1860 par MM. Pasteur, Cahours*. Reprinted in *Sur la dissymétrie moléculaire*, ed. L. Pasteur, J. H. van't Hoff, and A. Werner. Paris: Christian Bourgois.

Pasteur, L. [1868] 1939. "Les laboratoires. Le budget de la science." *Revue des cours scientifiques*, 5, pp. 137–139. Reprinted in *Oeuvres*, vol. 7, pp. 199–204. Paris: Masson.

Perrin, J. [1913] 1970. *Les Atomes*. Paris: Gallimard.

Pignarre, P. 1990. *Ces drôles de médicaments*. Paris: Edition des Laboratoires Delagrange.

Porter, T. M. 1981. "The Promotion of Mining and the Advancement of Science: The Chemical Revolution of Mineralogy." *Annals of Science,* 38, pp. 543–570.

Rappaport, R. 1961. "Rouelle and Stahl: The Phlogistic Revolution in France." *Chymia,* 7, pp. 73–101.

Redondi, P. 1987. Galileo Heretic, trans. R. Rosenthal. Princeton, NJ: Princeton University Press.

Reuben, B. G., and M. L. Burstall. 1973. *The Chemical Economy: A Guide to the Technology and Economics of the Chemical Industry.* London: Longman.

Rey, A. 1923. *La Théorie de la physique chez les physiciens contemporains.* Paris: Alcan.

Rocke, A. J. 1984. *Chemical Atomism in the Nineteenth Century: From Dalton to Cannizzaro.* Columbus: Ohio State University Press.

———— 1993. *The Quiet Revolution: Hermann Kolbe and the Science of Organic Chemistry.* Berkeley: University of California Press.

Rossiter, M. W. 1975. *The Emergence of Agricultural Science: Justus Liebig and the Americans, 1840–1880.* New Haven: Yale University Press.

Roth, E. 1988. "Highlights in the History of Analytical Chemistry in France." *Euroanalysis VI,* ed. E. Roth, pp. 1–27. Les Ulis: Editions de la Physique.

Russell, C. A. 1971. *The History of Valency.* Leicester: Leicester University Press.

———— 1985. *Recent Developments in the History of Chemistry.* London: Royal Society of Chemistry.

———— 1987. "The Changing Role of Synthesis." *Ambix,* 4, pp. 169–180.

———— 1988. "Presidential Address: Rude and Disgraceful Beginnings—A View of the History of Chemistry from the Nineteenth-Century." *British Journal for the History of Science,* 21, pp. 273–294.

Russell, C. A., N. G. Coley, and G. K. Roberts. 1977. *Chemists by Profession.* Milton Keynes: Open University Press.

Sadoun-Goupil, M. 1977. *Le Chimiste C. L. Berthollet 1748–1822, sa vie, son œuvre.* Paris: Vrin.

Shapin, S., and S. Schaffer. 1986. *Leviathan and the Air-Pump: Hobbes, Boyle and the Politics of Experiment.* Princeton, NJ: Princeton University Press.

Shea, W. R., ed. 1988. *Revolutions in Science: Their Meaning and Relevance.* Canton, MA: Science History Publications.

Sheppard, H. J. 1970. "Alchemy: Origin or Origins?" *Ambix,* 17, pp. 69–84.

Shinn, T. 1979. "The French Faculty System, 1808–1914." *Historical Studies in the Physical Sciences,* 10, pp. 271–233.

Smith, J. G. 1979. *The Origins and Early Development of the Heavy Chemical Industry in France.* Oxford: Clarendon Press.

Soddy, F. 1926. *Le Radium. Interprétations et enseignements de la radioactivité.* Paris: Félix Alcan.

Spronsen, J. W. van. 1969. *The Periodic System of Chemical Elements: A History of the First Hundred Years.* Amsterdam: Elsevier.

Stengers, I. 1989. "L'affinité ambiguë: le rêve newtonien de la chimie du XVIIIe siècle." In *Eléments d'histoire des sciences,* ed. M. Serres, pp. 297–320. Paris: Bordas.

Sutton, M. A. 1976. "Spectroscopy and the Chemists." *Ambix,* 23, pp. 16–26.

Svabadvary, F. 1966. *History of Analytical Chemistry.* Oxford: Pergamon Press.

———— 1979. "Early Laboratory Instruction." *Journal of Chemical Education,* 56, p. 794.

Thackray, A. W. 1966. "The Emergence of Dalton's Chemical Atomic Theory." *British Journal for the History of Science,* 3, pp. 1–23.

———— 1970. *Atoms and Powers: An Essay on Newtonian Matter-Theory and the Development of Chemistry.* Cambridge, MA: Harvard University Press.

Thackray, A. W., J. L. Sturchio, P. T. Caroll, and R. F. Bud. 1985. *Chemistry in America, 1876–1976: Historical Indicators.* Dordrecht: Reidel.

Thénard, L. 1813–1816. *Traité de chimie élémentaire théorique et. pratique,* 4 vols. Paris: Librairie Crochard.

Thomson, T. 1804. *A System of Chemistry,* 2nd ed., 4 vols. Edinburgh: Bell and Bradfute.

———— 1830–1831. *The History of Chemistry,* 2 vols. London.

Thorpe, E. 1902. *Essay in Historical Chemistry.* London: Macmillan.

Travis, A. S. 1993. *The Rainbow Makers: The Origins of the Synthetic Dyestuffs Industry in Western Europe.* Bethlehem, PA: Lehigh University Press.

Tren, T. 1979. "Rutherford's Radioactivity and α-Ray Research." *Ambix,* 21, pp. 53–77.

Turner, R. S. 1971. "Justus Liebig versus Prussian Chemistry: Reflections on Early Institute-Building in Germany." *Historical Studies in Physical Sciences,* 3, pp. 137–182.

van't Hoff, J. H. [1887] 1986. *La Chimie dans l'espace. Dix années dans l'histoire d'une théorie.* Rotterdam. Reprinted in *Sur la dissymétrie moléculaire,* ed. L. Pasteur, J. H. van't Hoff, and A. Werner. Paris: Christian Bourgois.

Veillerette, F. 1987. *Philippe Lebon ou "l'homme aux mains de lumière."* Paris: N. Mourot.

Vène, J. 1976. *Les Plastiques.* Paris: Presses universitaires de France.

Venel, G. F. 1753. "Chymie." In vol. 3 of the *Encyclopédie.* Paris.

Vidal, B. 1989. *Histoire de la liaison chimique: le concept et son histoire.* Paris: Vrin.

Von Baeyer, H. D. 1993. *Taming the Atom.* New York: Random.

Warren, K. 1980. *Chemical Foundations: The Alkali Industry in Britain to 1926.* Oxford: Clarendon Press.

Wehefritz, V. 1987. "Bibliography of the History of Chemistry." *Iatul Quarterly,* 1, pp. 162–167.

Weisz, G. 1979. "The French Universities and Education for the New Professions, 1885–1914: An Episode in the French University Reform." *Minerva,* 17, pp. 98–128.

Westfall, R. 1972. "Newton and the Hermetic Tradition." In *Science, Medicine and Society in the Renaissance: Essays to Honor W. Pagel,* ed. A. G. Debus. London: Heinemann.

———— 1975. "The Role of Alchemy in Newton's Career." In *Reason, Experiment and Mysticism,* ed. M. L. Righini Bonelli and W. R. Shea. London: Macmillan.

———— 1981. *Never at Rest.* Cambridge: Cambridge University Press.

Weyer, J. 1976. "The Image of Alchemy in Nineteenth and Twentieth Century Histories of Chemistry." *Ambix,* 23, pp. 65–79.

Wojtkowiak, B. 1988. *Histoire de la chimie.* Paris: Technique et Documentation-Lavoisier.

Wotiz, J. H., R.S. 1987. "The Unknown Kekulé." In *Essays on the History of Organic Chemistry,* ed. J. G. Traynham, pp. 21–34. Baton Rouge: Louisiana State University Press.

Wurtz, A. 1869. *Histoire des doctrines chimiques.* Paris: Hachette.

———— 1879. *La Théorie atomique,* 2nd ed. Paris: Alcan.

INDEX OF NAMES